GEOMETRIC MAGIC SQUARES

A Challenging New Twist Using Colored Shapes Instead of Numbers

LEE C. F. SALLOWS

DOVER PUBLICATIONS, INC.
Mineola, New York

Bibliographical Note
Geometric Magic Squares is a new work,
first published by Dover Publications, Inc., in 2013.

International Standard Book Number
ISBN-13: 978-0-486-48909-4
ISBN-10: 0-486-48909-4

Manufactured in the United States by Courier Corporation
48909401
www.doverpublications.com

Contents

For Evie,
Alright Rat?

Note: The numbers in square brackets in the text refer to the References at the end of the book.

Foreword

It is now almost forty years since a rainy afternoon in Nijmegen when I found that a fit of absent-minded doodling had inadvertantly developed into an attempt to construct a *magic square.* I was twenty-six at the time, a Brit recently arrived in Holland, where I was destined to remain domiciled down to the present day. Nijmegen is a town on the eastern side of the Netherlands, not far from Arnhem, close to the border with Germany. The reasons that had brought me to Nijmegen might be of interest in a different context, but are of no relevance to the present account. Like the First Patriarch of Zen Buddhism, I had just blown in from the West. The year was 1970, and I had landed a job as an electronics engineer at the local university, getting in via the back door as a member of their non-academic staff. The lab was new and set amid woodland. Compared to the working conditions I'd known back home, this was paradise. Also, the technical equipment available in the department was years ahead of anything I'd seen in England. Moreover, my income had doubled. Small wonder then that I stayed, although little did I then realize that it would be for life.

I cannot now even recall how or where I had first learned what a magic square is. The most probable source would have been one of the comics I read as a boy growing up in London, most likely my favourite, *The Eagle,* which enlivened its pages with verbal and mathematical curiosities, and had introduced me to my first palindrome, "Evil rats on no star live." A palindrome is a sentence that reads the same backwards as it does forwards. My own name can be appended backwards to itself to produce the almost-palindrome, "Lee Sallows swollas eel." If you like palindromes then there is a good chance you will like magic squares too. It is a thirst for symmetry shared by many. In any case, the concept of a magic square had somehow lodged in my mind, so that the purpose guiding my doodling hand all those years later was at least clear. It was to arrange the numbers from 1 to 9 in a square grid of nine cells so that the sum of any three of them lying in a straight line would be the same. Eight such straight-line-triplets can be traced in a square of nine cells, three of them forming its rows, three its columns, and two its corner-to-corner diagonals. Warming to my task, it seemed to me sufficiently wonderful that nine numbers could be found to realize the desired result at all, let alone that it could be achieved using the natural numbers from one to nine. However, I could clearly remember having seen such a square somewhere before, so there was no doubt in my mind that it could be done. In fact, the puzzle presents little difficulty, and within a few minutes I was able to examine my solution at leisure. It can be seen in Figure 1.1 of the Introduction, the very first illustration in this book.

To the receptive eye there is something deeply satisfying in such a square. See, for example, how the four pairs straddling the centre number each sum to ten. There are just four possible ways to split ten into two distinct whole numbers, and all of them appear here. Similarly, there are just eight possible ways in which three different decimal digits can combine to yield 15. Again, every one of them is to be found occupying a row, column, or diagonal in the magic square.

In my mind's eye, I imagined the numbers replaced by weights of 1, 2, 3, . . ., standing on a square board at nine spots corresponding to the exact centre of each cell in the 3×3 grid. Ideally, these weights would all be of the same size and shape. Underneath the board, at its exact centre, was a pivot. And upon this pivot the board would perch in perfect balance, a reflection of the numerical balance realized in the magic square. Years later I saw a photograph of an almost identical construction due to Craig Knecht, only with the board suspended from its centre by a wire, instead of balanced on a pivot[1]. Craig had the clever idea of building his weights from different numbers of identical metal rings, or washers, piled atop each other on vertical posts that stuck upwards from the centre of every square cell.

But back to that rainy day in Nijmegen. Re-examining my new plaything, questions began to form. Could a second magic square be produced by arranging the same numbers differently? What of alternative numbers? Simple and pleasing as are 1,2,3, . . . , clearly others might be used to similar effect. Then again, were larger magic squares to be found? Could the numbers from 1 to 16 be used to

1 Knecht's construction can be viewed on Harvey Heinz's wonderfully rich website devoted to magic squares: http://www.magic-squares.net.

produce a 4×4 specimen? Let's see, supposing we call the number in the top left-hand cell . . . I had no training in mathematics, but I found myself trying to recall some of the half-forgotten algebra learned at school. Little did I know it, but I was playing with fire. There are some people for whom magic squares are more addictive than cocaine. I should know; here I am forty years later and still hooked.

It was soon after this that I first ran across Martin Gardner's "Mathematical Games" column in *Scientific American,* and was thus introduced to the world of recreational mathematics. With his genius for making difficult things easy, Gardner led me into that world until I felt confident enough to start making my own way. There was no Internet in those days. It was only through Martin Gardner's writings that I became aware of the extensive literature on magic squares, a literature that I now began to explore. However, beyond this there were the rows of bound volumes of *Scientific American* I found lining the shelves of our university library. The series went back some twenty-five years, and every single issue contained an as-yet-to-be-read Mathematical Games column. It took me a around a month to work through the entire series, an exercise from which I emerged with more or less glazed eyes and a burning desire to contribute to the field something new of my own. Gardner had woken me up to the fact that many of the ideas and inventions found in Mathematical Games were the products of *amateurs;* that you didn't need a diploma to do your own research, that permission is not required before making a new discovery.

It was, in fact, in the field of magic squares that I landed my first ever new 'find,' which was an improved algebraic generalization of 4×4 squares. A trifling innovation, it was nonetheless poignant for me in showing that it was indeed possible for a rank amateur to make an original discovery, an insight from which I never looked back. Thus, while no mathematical mountaineer, I set out on my own to explore low-level routes, discovering in the process a great liking for mathematical wordplay. It was in this way that what had become a compulsive habit of playing with words and numbers lead eventually to the idea of 'alphamagic' squares, a frivolous invention that has come to enjoy a modest renown. Some examples can be found in this volume. Still, later, these were followed by two more novelties in the shape of 'septivigesimal' magic squares and 'ambimagic' squares. *Septivigesimal* is just a long word for "twenty-seven-based." It refers to a certain *gematria* or system for interpreting words as numbers. The magic squares referred to are ones in which *words* occupy the cells, their values, as interpreted in the gematria, then emerging as numbers that form a conventional magic square. Ambimagic squares come up for brief elucidation in Part II of this book. Of course, magic squares were only one among other topics in recreational math to attract my interest. But that they have loomed large in my thoughts over a long period of years cannot be denied.

I first hit on the idea of a *geometric* magic square in October 2001, and I sensed at once that I had penetrated some previously hidden portal and was now standing on the threshold of a great adventure. It was going to be like exploring Aladdin's Cave. That there were treasures in the cave, I was convinced, but how they were to be found was far from clear. The concept of a geometric magic square is so simple that a child will grasp it in a single glance. Ask a mathematician to create an actual specimen and you may have a long wait before getting a response; such are the formidable difficulties confronting the would-be constructor.

The present volume seeks to throw some light on these matters. Although only a feeble light, it must be admitted. For the truth is that the informal methods here introduced are a far cry from what I would have liked to present, given the mathematical character of the topic. I can only say that a more rigorous treatment has proved beyond my amateur capacities. As a magic-square enthusiast, my guiding instincts have been those of the collector, whose chief aim is ever to add a new specimen to his collection. It is an ambition that is unlikely to go hand-in-hand with mathematical rigour. But if you are looking for a hunter's handbook that includes tips on how to set traps and stalk prey, and so on, then perhaps you could do worse.

This book has been long in the making, the writing having at one stage been abandoned in favour of creating a website devoted to the same topic: http://www.geomagicsquares.com. But life can play funny tricks. From the start, the website turned out to attract more attention than was ever anticipated. Soon after launching, the journal *New Scientist* ran an enthusiastic review of the site by Alex Bellos, in the wake of which 14,000 hits were received on one day. Still later, Michael Kleber offered to reproduce the website Introduction in his "Mathematical Entertainments" column in *The Mathematical Intelligencer.* It appeared in the Winter edition for 2011, Vol. 23, No. 4, the cover of which was adorned with a geomagic square. A lively interest in the site continues down to the present day. It was thus largely in response to this surprising success that I was prompted to resume writing and so bring this book to completion. In the appendices I have included some further material on magic squares that I hope may prove of interest. There is a good deal of overlap among these pieces which were written at various periods over the past thirty years. However, they include some personal insights that are not to be found elsewhere, and they will paint a picture of the way my thoughts were turning in the years preceeding the emergence of geometric magic squares.

Lee Sallows
Nijmegen, September 2011

Part I
Geomagic Squares of 3x3

There is mystery in symmetry. With an m to spare.

1 Introduction

I expect most readers will be familiar with the traditional *magic square:* a chessboard-like array of cells in which numbers, usually but not always consecutive, are written so that their totals taken in any row, any column, or along either diagonal, are alike. Figure 1.1 shows the best known example of size 3×3, the smallest possible, a square of Chinese origin known as the *Lo shu.* The constant sum of any three entries in a straight line is 15. The diagram at left shows the *Lo shu* in traditional form, an engaging device nowadays identified by sinologists as a pseudo-archaic invention of the 10th century A.D.; see Cammann[1]. The dot-and-line notation was intended to suggest an origin of extreme antiquity.

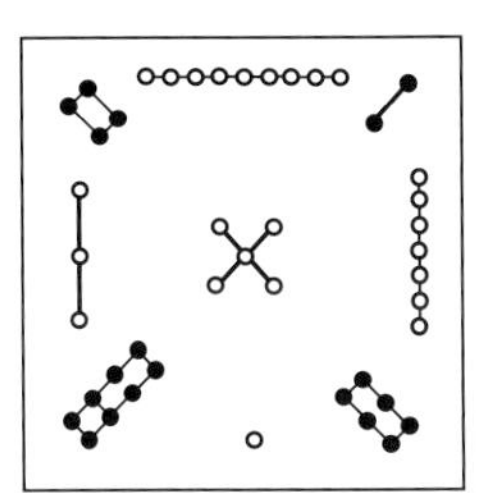

4	9	2
3	5	7
8	1	6

Fig. 1.1 The *Lo shu.*

The history of magic squares is a venerable one, earliest writings on the topic dating from the 4th century BC[2]. Abstruse as they may appear, these curiosities have long exercised a peculiar fascination over certain minds, attracting over the centuries a steady following of devotees, by no means all of them mathematicians. As Martin Gardner has written, "The literature on magic squares in general is vast, and most of it was written by laymen who became hooked on the elegant symmetries of these interlocking number patterns."

It is true. I myself am such a layman; a mathematical amateur with an irrational fondness for the crystalline quality of these numerical prisms (see, for example, [3] and [4].) But with that humble position owned up to, my purpose in the present essay is in fact decidedly less timid.

My thesis is that the magic square is, and has ever been, a *misconstrued* entity; that for all its long history, and for all its vast literature, it has remained steadfastly unrecognised for the essentially *non-numerical* object it really is. Just as a cylinder may be mistaken for a circle when observed from a single viewpoint, so may a familiar object turn into something quite unexpected when seen from a new perspective. In a similar vein, I suggest the numbers that appear in magic squares are better understood as symbols standing for (degenerate instances of) geometrical figures. Hence the prefix *geometric* to distinguish the wider genus of magic square that will turn out to include the old species within it. For, as we shall see, the traditional magic square is really no more than that special instance of a geometric magic square in which the entries happen to be *one-dimensional.* But once we are introduced to squares using two-dimensional entries the scales fall from our eyes and we step into a wider, more exhilarating world in which the ordinary magic square occupies but a humble position.

2 Geometric Magic Squares

Consider a graphical representation of the *Lo shu* as seen in Figure 2.1 at right, in which straight line segments of length 1, 2 ,3, . . . replace like-valued numbers in each cell. The orientation of these segments in their cells is unimportant; they may be horizontal, vertical, or slanted at any angle. The constant sum, 15, as represented by 8+1+6 in the bottom row, say, then becomes three segments of length 8, 1, and 6 that are joined head to tail so as to form a single straight line of length 15.

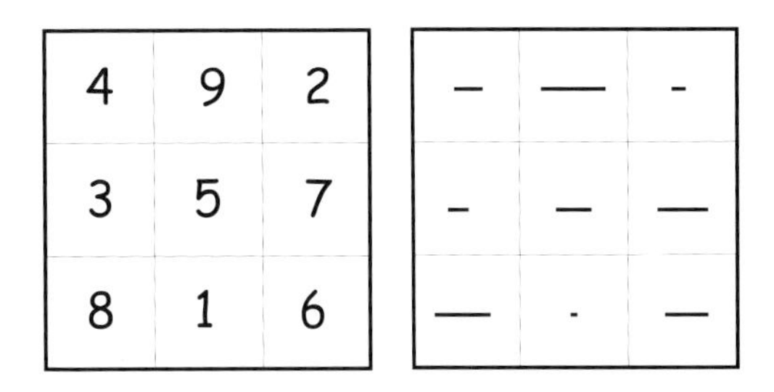

Fig. 2.1 A geometrical version of the *Lo shu.*

We note that the order in which 8, 1, and 6 are abutted is non-critical, the important thing being only that they fit together end-to-end so as to fill or 'pave' a straight line of length 15. And similarly for the seven other sets of three line segments occupying the remaining rows, columns, and main diagonals, collectively known as 'magic lines.' Hence more generally:

(1) The numbers that occur in magic squares can be seen as abbreviations for their geometrical counterparts, which are *straight line-segments* of appropriate length.

(2) The process of adding numbers so as to yield the recurring constant sum is then easier to interpret as the arithmetical counterpart of *partitioning* or *tiling a space* with these line segments.

The advantage of this view now emerges in an entirely novel contingency it immediately suggests. For just as line segments can pave longer segments, so *areas* can pave larger areas, *volumes* can pack roomier volumes, and so on up through higher dimensions. In traditional magic squares, we add numbers so as to form a constant sum, which is to say, we 'pave' a one-dimensional space with one-dimensional 'tiles.' What happens beyond the one dimensional case?

Geometric or, less formally, *geomagic* is the term I use for a magic square in which higher dimensional geometrical shapes (or *tiles* or *pieces*) may appear in the cells instead of numbers. For the moment we shall dwell on flat, or two-dimensional shapes, although non-planar figures of 3 or higher dimensions may equally be used. The orientation of each shape within its cell is unimportant. Such an array of $N \times N$ geometrical pieces is called *magic* when the N entries occurring in each row, each column, as well as in both main diagonals, can be fitted together jigsaw-wise to produce an *identical shape* in each case. In tessellating this constant region or *target*, pieces are allowed to be flipped. As with numerical, or what I now call *numagic* squares, geomagic squares showing repeated entries are denoted (and deemed) *trivial* or *degenerate*, which are terms we shall have need of more often. Rotated or reflected versions of the same geomagic square are counted identical, as are rotations and reflections of the target. A square of size $N \times N$ is said to be of *order N*. We say that a geomagic square is of *dimension D* when its constituent pieces are all D-dimensional. This is an informal introduction to geomagic squares; for a formal definition see Appendix I. In the following, our concern will be almost exclusively with 2-D, or two-dimensional squares.

Figure 2.2 shows a 3×3 two-dimensional geomagic square in which the target is itself a square. Any 3 entries

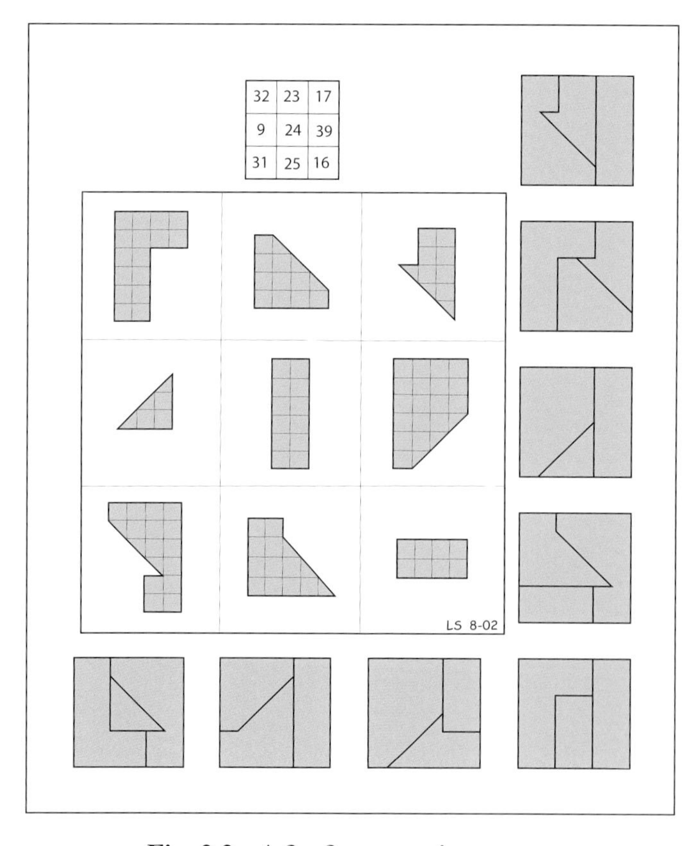

Fig. 2.2 A 3 × 3 geomagic square.

in a straight line can be assembled to pave this same square-shape without gaps or overlaps, as illustrated to right and below. Note how some pieces appear one way in one target, while flipped and/or rotated in another. Thin grid lines on pieces within the square help identify their precise shape and relative size.

At the top is a smaller 3×3 square with numbers indicating the areas of corresponding pieces in the geomagic square, expressed in units of half grid-squares. Since the three pieces in each row, column, and diagonal tile the same shape, the sum of their areas must be the same. This is, therefore, an ordinary numagic square (or one-dimensional geomagic square) with a constant sum equal to the target area. Analogous area squares for many geomagic squares are often degenerate because differently shaped pieces may share equal areas.

The concept of geometric magic squares grew out of an original impulse to create a pictorial representation of the algebraic square shown in Figure 2.3, a formula due to the 19th century French mathematician Édouard Lucas[5] that describes the structure of every 3×3 numagic square. The *Lo shu*, for example, is that instance of the formula in which $a = 3$, $b = 1$, and $c = 5$. From here on the terms *formula* and *generalization* will be used interchangeably.

The idea underlying this pictorial representation was as follows. Suppose the three variables in the formula are each represented by a distinct planar shape. Then the entry $c + a$ could be shown as shape c *appended* to shape a, while the entry $c - a$ would become shape c from which shape a has been *excised*. And so on for the remaining entries. A back-of-the-envelope trial then lead to Figure 2.4, in which a is a rectangle, b a semi-circle, and c a (relatively larger) square, three essentially arbitrary choices.

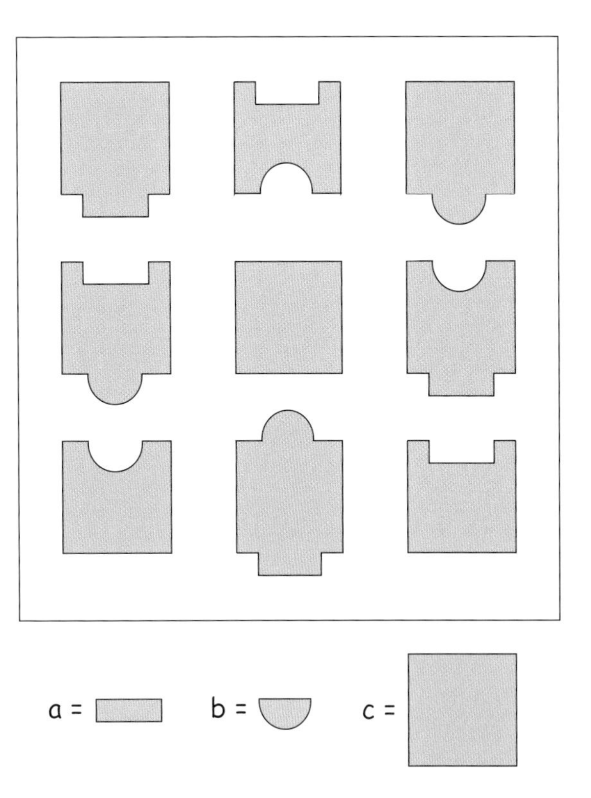

Fig. 2.4 Lucas's formula realised in geometric shapes.

$c+a$	$c-a-b$	$c+b$
$c-a+b$	c	$c+a-b$
$c-b$	$c+a+b$	$c-a$

Fig. 2.3 Lucas's formula for the general 3x3 numagic square.

This result was more effective than anticipated, the match between protrusions and indentations ("keys" and "keyholes") making it easy to imagine the pieces interlocking, and thus visually obvious that the total area of any three in a straight line is the same as a rectangle of size 1×3, or three times the area of the central piece, in agreement with the formula. However, the fact that the 3 central row and 3 central column pieces will not actually fit together to complete a rectangle, as the pieces in all other cases will, now seemed a glaring flaw. The desire to find a similar square lacking this defect was then inevitable, and the idea of a geometric magic square was born.

It was not until later, however, that the relationship between geometric and traditional magic squares became clear. For, as we have seen, although the term *geometric magic square* may seem to suggest a certain kind of magic square, in fact things are the other way around. On the contrary, it is ordinary magic squares that turn out to be a special kind of geometric magic square, the kind that use one-dimensional pieces.

The problem of how to actually produce such a square now took centre stage. Following a lot of thought on this question, thus far two approaches have suggested themselves: (1) pencil and paper methods based on algebraic formulae, along the lines just mentioned. (2) in the case of squares restricted to *polyforms* or shapes built up from repeated atoms, brute force searches by computer. For short, I call the latter *polymagic* squares, which is probably a misnomer, but no matter. Foremost among the polyforms are polyominoes (built up from unit squares), polyiamonds (equilateral triangles) and polyhexes (regular hexagons). Figure 2.5 shows 'Magic Potion', an example using nine hexominoes. I'm afraid I have been unable to resist the temptation of assigning titles to some of the better specimens. In searching for such a square different target shapes must be tried. In this case, the result was felicitous. In general, both

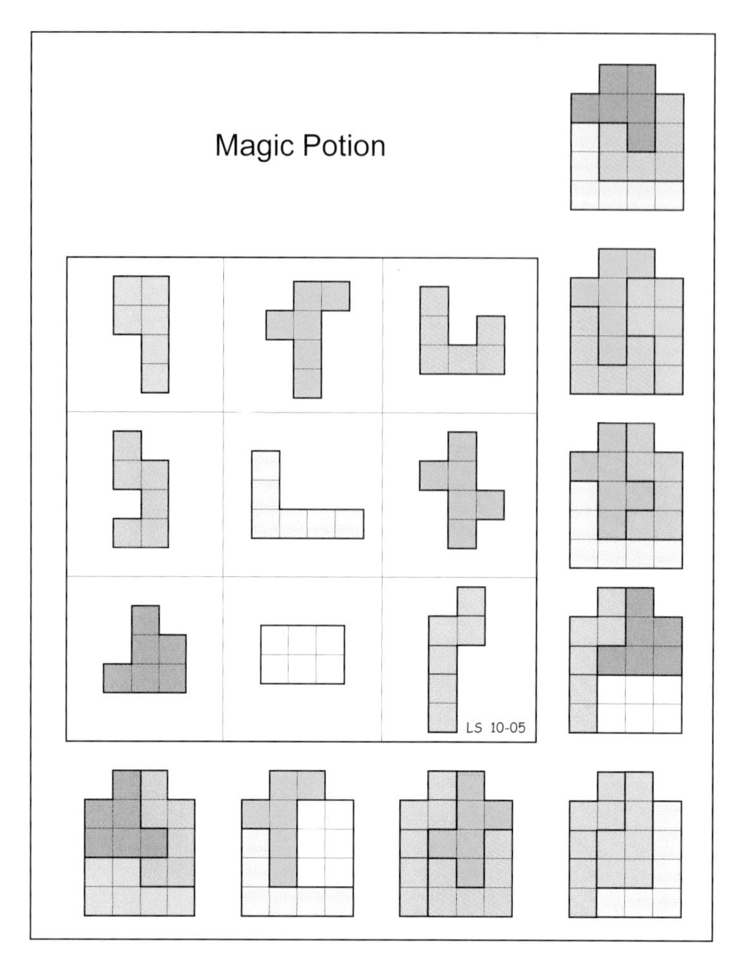

Fig. 2.5 'Magic Potion,' a polymagic square.

construction methods have proved fruitful. Some simple inferences that follow from Lucas's formula are an essential basis for both.

3 The Five Types of 3×3 Area Squares

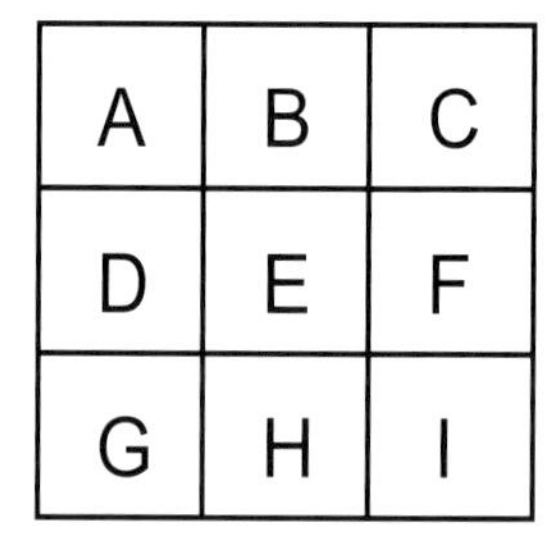

Fig. 3.1 Standard labelling for 3 × 3 squares.

Let G be a geomagic square, where G' is the square that results from replacing each piece in G by a number representing its area. Then, by the definition of a geomagic square, G' is a numagic square, although perhaps *degenerate,* since piece areas may repeat. Thus, if G' is of order-3, its entries must satisfy the relations expressed in Lucas's formula, and, if G is a polymagic square, these entries will be whole numbers. It is easily verified that these are distinct if, and only if, $a \neq \pm b$ or $\pm 2b$, or 0.

Consider now the possible forms that a degenerate magic square may take. We shall use Figure 3.1 as our standard for identifying cells. Suppose now that Figure 3.1 is a trivial square in which $A = B$.

Then by Lucas's formula, $c + a = c - a - b$, or $b = -2a$, which on substitution in the formula yields the type 1 square of Figure 3.2. Here we see the full set of relations implied by $A = B$, or equivalently, of $B = C$, $C = F$, $F = I$, $I = H$, $H = G$, $G = D$, and $D = A$, when rotations and reflections of the same square are in turn considered. Repeating this process for the cases $A = C$, $A = E$, . . . , $B = D$, . . etc, we discover just three further possible forms of a degenerate square, as seen in the remaining instances of Figure 3.2.

Thus, for every geomagic square G, either G' is a magic square in which every number is distinct or non-degenerate

type 1

c+a	c+a	c−2a
c−3a	c	c+3a
c+2a	c−a	c−a

type 2

c+a	c−2a	c+a
c	c	c
c−a	c+2a	c−a

c+a	c−a	c
c−a	c	c+a
c	c+a	c−a

type 3

c	c	c
c	c	c
c	c	c

type 4

Fig. 3.2 The four degenerate types of magic square. Note that a type t square contains exactly 9 - 2*t* different entries.

(call this type 0: Lucas's formula with $a \neq \pm b$ or $\pm 2b$ or 0), or G' is a degenerate magic square showing one of the four structures of Figure 3.2. We are now ready for a look at the first method for producing geomagics.

4 Construction by Formula

As discussed previously, every numerical magic square has a primitive geometrical analog using straight line-segments. We have only to broaden these lines into strips or rectangles of same height to result in a two-dimensional geomagic square, the target then being a longer strip that is formed simply by concatenating the shorter ones occupying each magic line (i.e., each row, column, and diagonal). By suitable choice of rectangle height, the target can even be made a square, as in the example based on the *Lo shu* shown in Figure 4.1.

Similarly, just as any set of contiguous points along the real number line can be mapped one-to-one onto another set of contiguous points around the circumference of a circle or part-circle, so numerical magic squares have another primitive geometrical analog using circular arcs or sectors of appropriate *angle*, the target then being the circle or part circle formed by subjoining these arcs or sectors. Figure 4.2 shows such a representation of the *Lo shu* using sectors. Since the constant sum in the *Lo shu* is 15, the smallest sector subtends an angle of $360 \div 15 = 24°$, the angles of the other sectors being multiples of 24°, up to $9 \times 24 = 216°$.

It is easy to see that this circular target could be replaced by a regular 15-gon, the sectors then changing to 15-gon segments of corresponding size. Likewise, the sectors in Figure 4.2 could be changed into annular segments, the target then becoming a ring with a central hole, or central 15-gon hole. Further variations may occur to the reader. By combining the straight line segment and circular arc interpretations, numerical squares could equally be mapped onto 3-*D helical* segments.

The rectangles and sectors in Figures 4.1 and 4.2 can be further elaborated. Earlier I spoke simplistically of 'broadening the line segments into strips of same height.' A better way of conceiving this is to think of the broadened strip as just two 1-*D* segments of same length, one above the other, their ends joined by two straight vertical lines

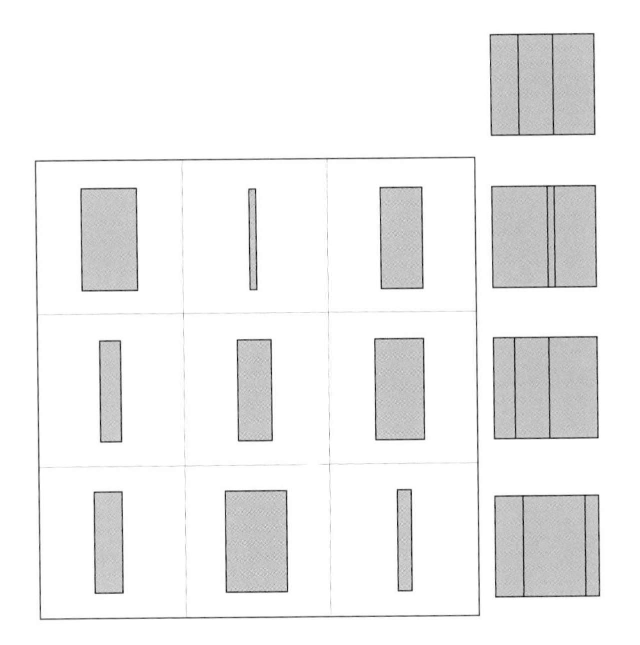

Fig. 4.1 Rectangles replace numbers in the *Lo shu*.

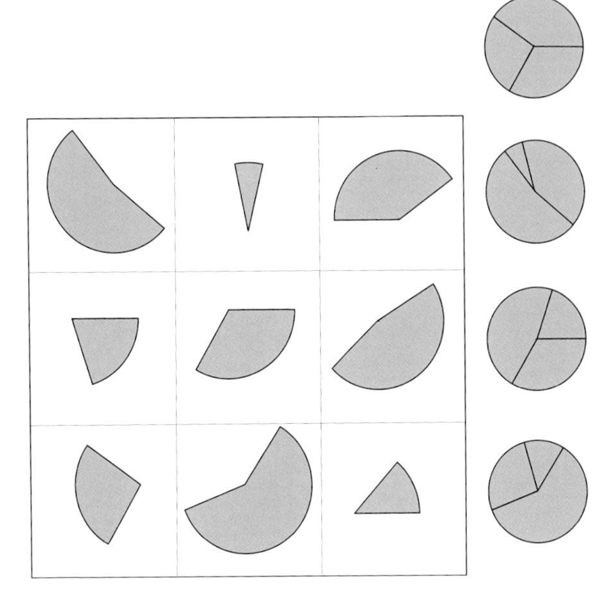

Fig. 4.2 Numbers in the *Lo shu* replaced by circular segments.

so as to form a rectangle. However, it is not necessary that these lines be straight, only that they be congruent. Imagine a piece formed by a pile of contiguous line segments, all parallel to each other, and yet shifted to left or right so that their ends describe some non-linear profile, as in Figure 4.3.

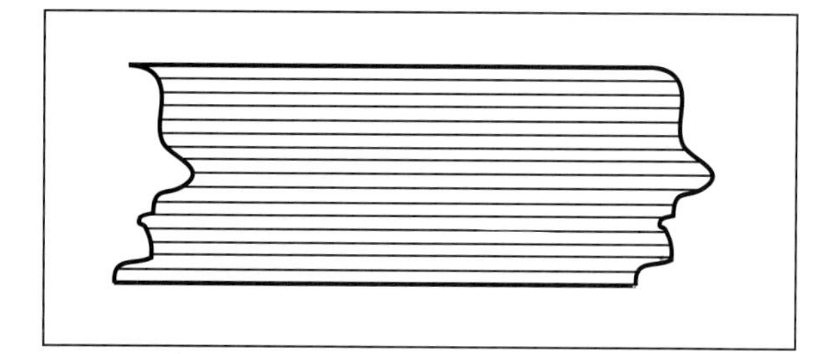

Fig. 4.3 Piece edges need only be congruent.

Provided all are contructed similarly, differing only in their lengths, pieces constructed in this way can again be concatenated to form a long, thin target whose ends are sculpted with the same curve. Similar remarks apply to circular segments, a striking example of the kind of profile just mentioned being realized in Figure 20.13 in the section on picture-preserving geomagic squares in Part 3.

This view of 2-*D* shapes as a stack of parallel straight line-segments appropriately aligned might seem to preclude shapes with re-entrant angles such as Figure 4.4 because the 1-*D* segments become broken. Happily, however, it turns out that this doesn't matter. In fact, it wouldn't matter if the projecting lug were entirely detached from the main body of the piece to become an island, so that that its corresponding indentation became an isolated hole. This brings us to *disconnected* pieces.

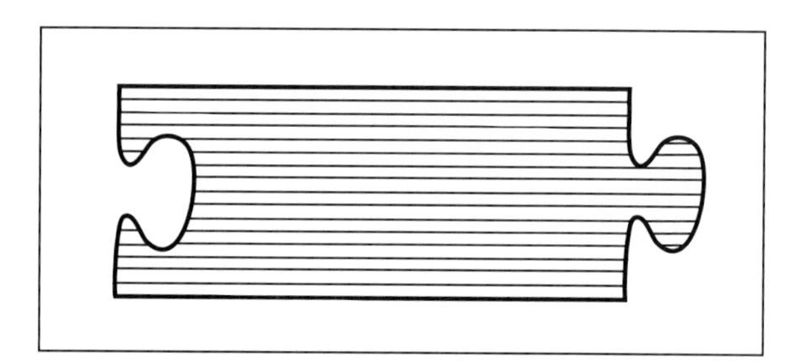

Fig. 4.4 A piece with broken line-segments.

Previously we saw that every numerical magic square corresponds to a 1-*D* geometrical magic square written in shorthand notation. But this is not to say that numerical squares account for *all* possible 1-*D* geomagic squares. In fact, they account only for that subset of 1-*D* squares using *connected* line segments. Figure 4.5 shows a 1-*D* geomagic square of order-3 that includes *disjoint* pieces, or pieces composed of two or more separated islands bearing a fixed spatial relation to each other. The overall shape of the compound piece is thus preserved even when moved. Here, the 1-*D* lines have been broadened and colored to enhance clarity, a trick that could obviously be extended so as to yield a true 2-*D* geomagic square sporting rectangular targets. However, the point to be made here is that Figure 4.5 is a 1-*D* geomagic square for which there exists no corresponding numerical magic square. Magic squares using numbers thus account for no more than a small fraction of all 1-*D* geomagic squares.

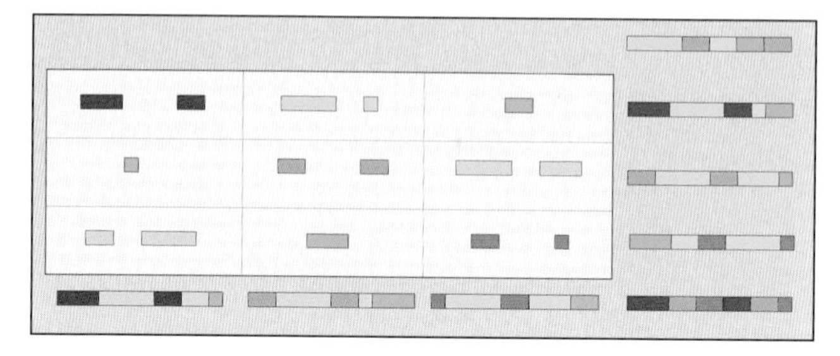

Fig. 4.5 A square using disconnected pieces.

Just as with linear pieces, so circular arc pieces do not have to be connected. Figure 4.6 shows a 3×3 square using disjoint arcs, their unit segments here simplified into single colored dots. Once again, such disconnected pieces cannot be represented by single numbers.

Of course, the trouble with geomagic squares of the

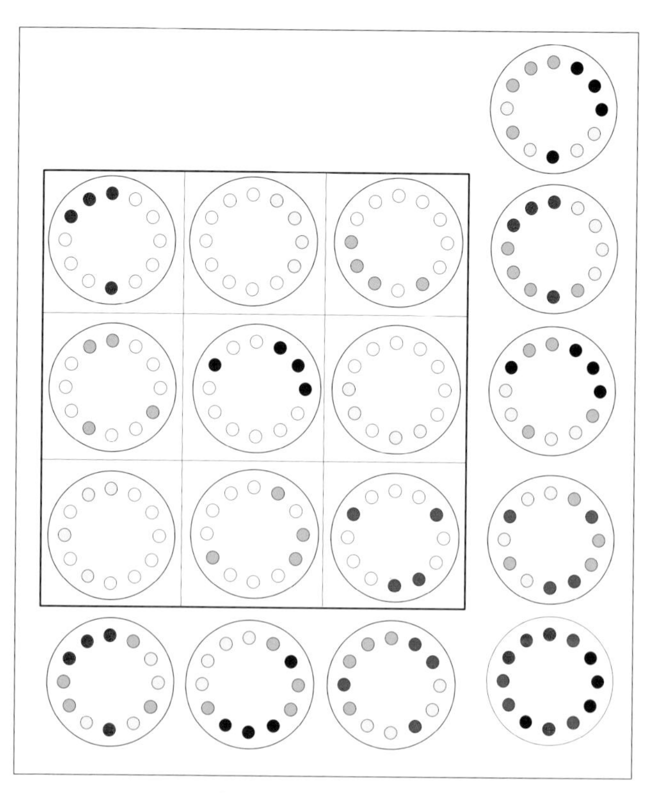

Fig. 4.6 Circular arcs may also be disconnected.

type seen in Figures 4.1 and 4.2 is that they are really nothing more than the same old numerical magic square in alternate guise. The question is: how do we go about producing more interesting 2-*D* geomagics such as the first one looked at in Figure 2.2, which are something other than just a geometrical rehash of an arithmetical square? One approach is to start with a trivial geomagic square based on a trivial algebraic formula, and then to *de-trivialise* this by adding appropriate keys and keyholes. I call algebraic squares, trivial or otherwise, that are used in this way, *templates*. An example will clarify.

Consider the trivial type 3 square (page 5), which is that case of Lucas's formula with $b = 0$. Setting $a = 1$ and $c = 2$, the lowest possible whole number values, results in Figure 4.7, a trivial numagic square with a constant line total or 'magic sum' of 6:

3	1	2
1	2	3
2	3	1

Fig 4.7 A type 3 numagic square.

As before, a primitive geometrical analog of this is easily produced, the magic sum of 6 now suggesting (on analogy with the 15-gon) a regular hexagon, say, as a nice choice of target. The latter can be divided radially into segments subtending angles of one, two, and three sixths of 360° so as to yield a triangle, parallellogram and trapezium respectively, as in Figure 4.8.

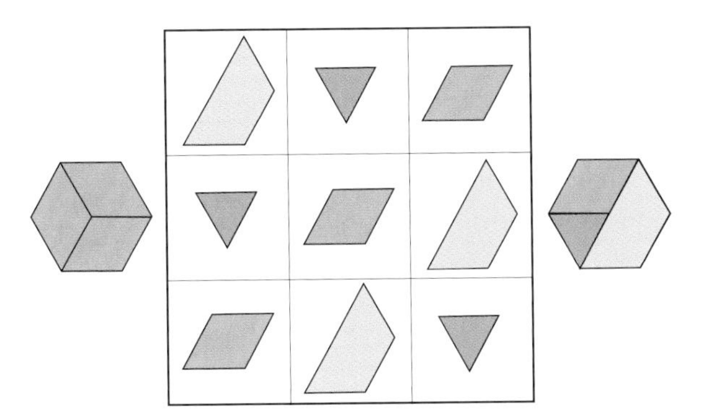

Fig 4.8 A substrate or trivial geomagic square.

The hexagons show the two ways in which the 3 pieces in each line assemble to complete the target. I call such an initial, necessarily trivial geomagic square that has yet to be elaborated, a *substrate*. Detrivializing this substrate so as to yield 9 distinct piece-shapes is then merely a matter of assigning a nominal shape (a small lug, say) to variable b in the type 3 formula, to result in the pattern of keys ($+b$) and keyholes ($-b$) seen in Figure 4.9.

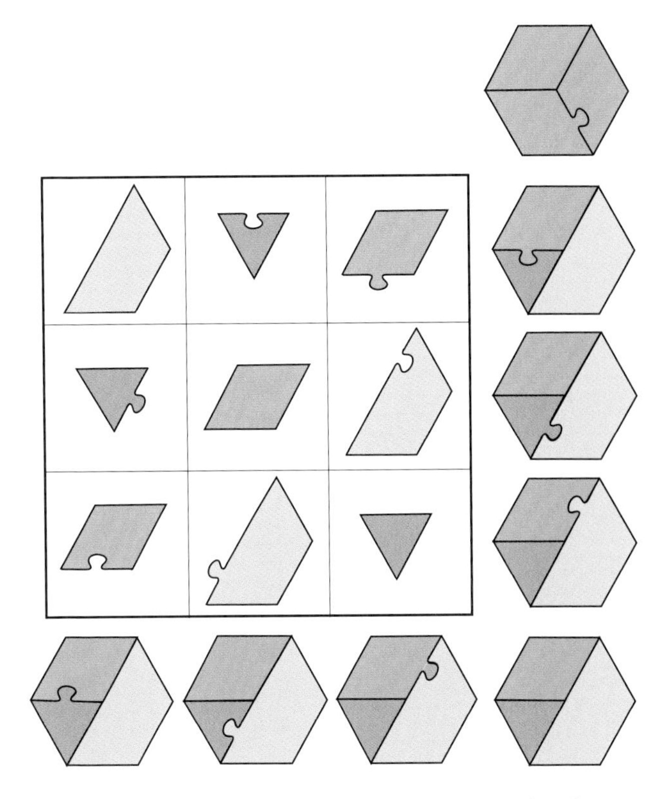

Fig 4.9 The trivial substrate detrivialised.

This is now a non-trivial 2-*D* geomagic square that is not simply a numerical magic square in different guise. Note that the keys and keyholes may be of *any* shape, provided only they remain within piece boundaries. They are thus in a sense *geometric variables*, in that they are arbitrary shapes that can be taken as standing for any other shape we might choose instead. The strong effect of an alternative choice of key/keyhole shape can be seen from *Magic Crystals* in Figure 4.10, which is a *polymagic* square, identical to Figure 4.9, except that the key shape is now a unit triangle belonging to the underlying isometric grid.

A striking result of this change is that the identity of the keys and keyholes as such now becomes lost to sight, making it far more difficult for the viewer to discern the principle of construction. So although Figure 4.10 is the prettier picture, as well as a greater feat of illusion, it is better understood as a particular instance of Figure 4.9, which provides the blueprint for an entire family of geomagic squares.

There are in fact two kinds of key/keyhole at work, in Figure 4.9; one obvious, the lug, the other invisible. It is again a triangle, the 60° hexagon segment that corresponds to variable a in Lucas's formula, and is half that of the 120° segment corresponding to variable c, the parallelogram. The effect of appending this a-shaped "key" to c is thus indistinguishable from enlarging c (the segment angle increases), and the effect of excizing the a-shaped key from c is indistinguishable from reducing c (the segment angle decreases). Similar effects are at work in Figure 4.1, where the rectangular "keys" and "keyholes" represented by both a and b merely increase or decrease the length of rectangle c. Call the latter "size-altering" keys/keyholes to distinguish them from "lug-type" keys/keyholes. Figure 4.11 shows another geometrical analog of Lucas's formula in which the variables a and b are now *both* represented by lug-type keys and keyholes.

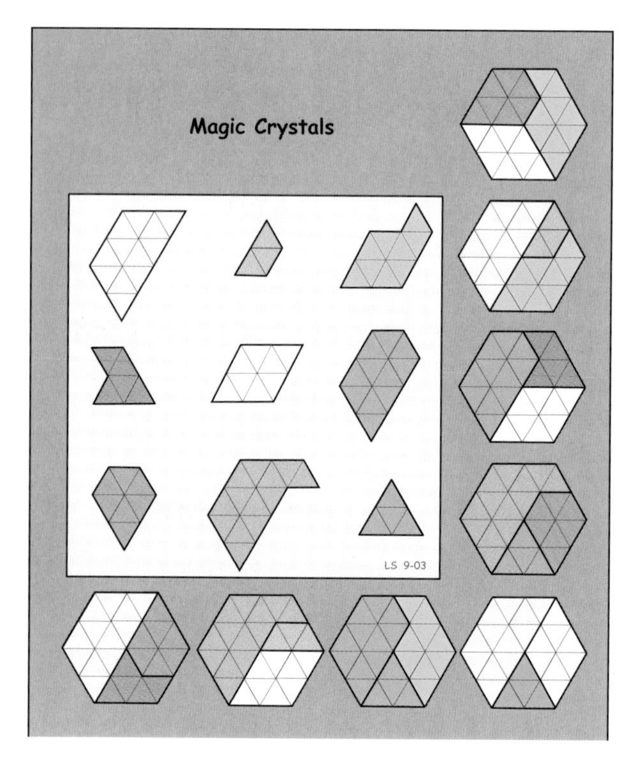

Fig 4.10 'Magic Crystals' The principle of construction is difficult to detect.

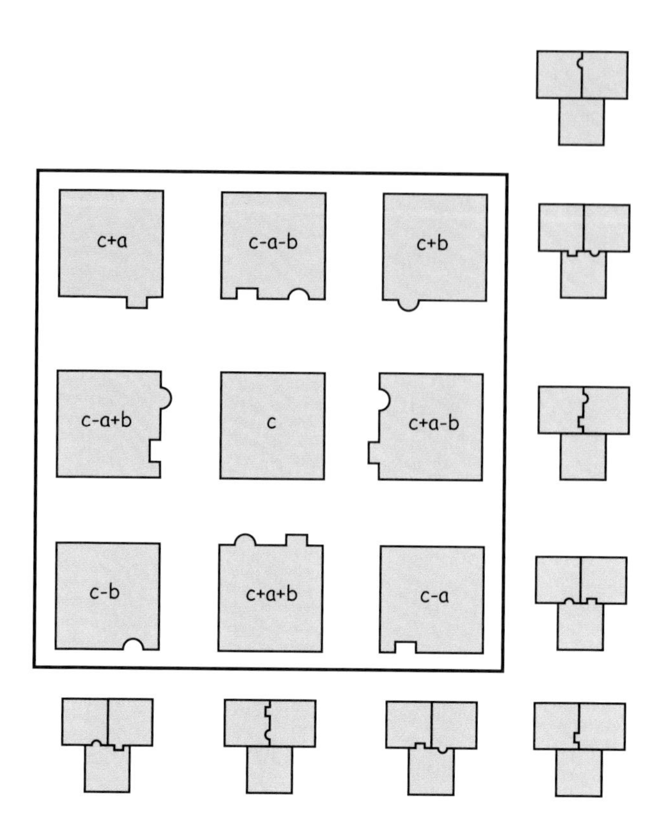

Fig. 4.11 Lucas's formula used as template.

Looking again at Lucas's formula, we see that the trivial square started with here is a uniform array with c in each cell, giving rise to a substrate composed of 9 identical pieces. It took a little while to arrive at the choice of 9 *squares*. My difficulty was in seeing how two keys on one piece, whatever its shape, could be made to marry with two keyholes on another piece of same shape (as for example in the centre column), *as well as* two single keyholes on two separate pieces (e.g. in the bottom row). The T-shaped target provided a solution, the only one I have been able to find. Figure 4.12 shows a different rendering of Figure 4.11 using polyominoes. Variable c is now a square 16-omino, a a square tetromino and b a domino. The target remains a T-shape. See again how the identity of the keys and keyholes disappears from view with the change to polyominoes.

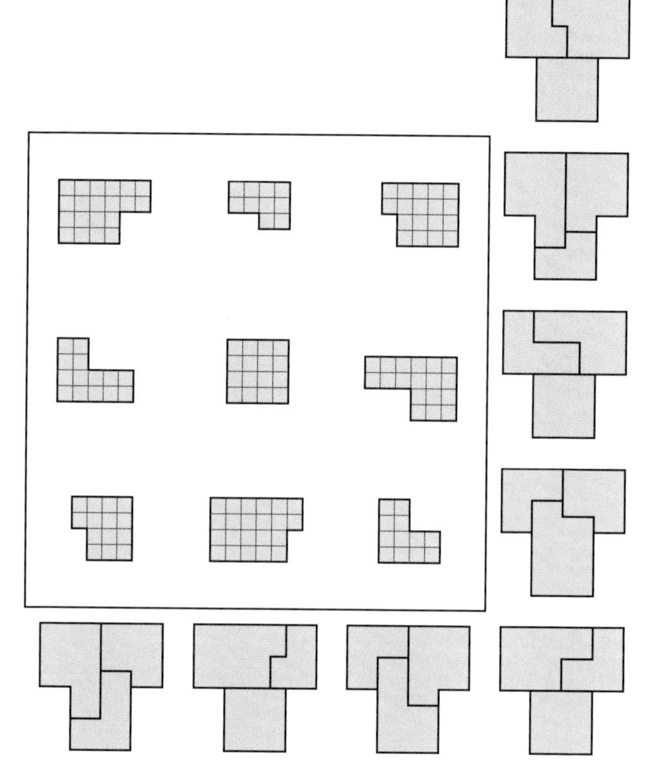

Fig. 4.12 Polyominoes obscure the keys and keyholes.

Compare this now with an alternative geometrical analog of Lucas's formula (Figure 4.13) in which variables *a* and *b* are of lug-type and size-altering type keys, respectively. The target here, not shown, is of course simply a rectangle of length 3 times the width of *c*.

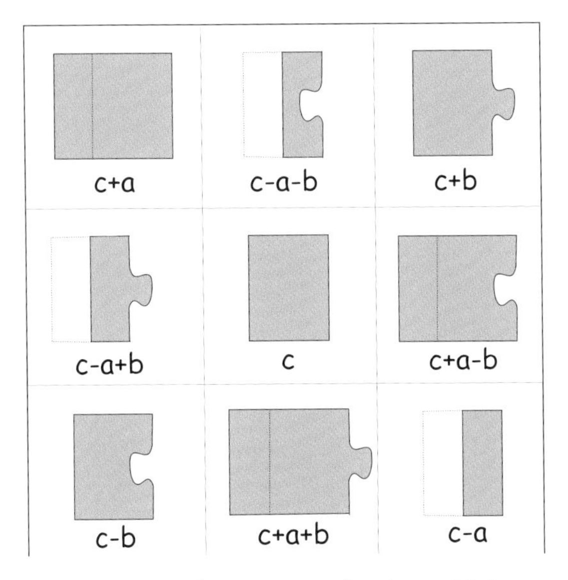

Fig. 4.13 Combining size-altering with lug-type keys and keyholes.

In the foregoing we have been considering a single example. However, the important point is that just as Figure 4.9 was derived from the trivial square of type 3 on page 5, so different piece schemes will emerge when we start from a different trivial type. Likewise, setting $c=2$ in the type 3 square gave a magic sum of 6 in Figure 4.7, which is the origin of the hexagonal target chosen. Different assignments will suggest different regular polygons as targets. Moreover, Figure 4.8 shows but one way to divide a hexagon into one, two, and three sixths of 360°; there are many others. Nor need the target be a regular hexagon, the essential requirement being in fact only *six-fold rotational symmetry*. Figure 4.14 shows an alternative to a hexagon in the form of a six-pointed star. Here, the circular presentation makes it a little harder for the viewer to spot which star belongs to which row, column or diagonal. Even so, it can be done without recourse to a telescope; Figure 4.15 provides a key.

So far so good, but at this point a word of caution. In section 3, we saw that the area square, *G'*, of a geomagic square, *G*, must always correspond to one of the five algebraic square types listed. Thereafter we took one of those types, type 3, and used it as a template to create the non-trivial square in Figure 4.9. But this doesn't mean that every 3×3 geomagic square can be identified as a particular instance of one of those types. For example, there is at least one further subtlety at work here about which we need to be aware.

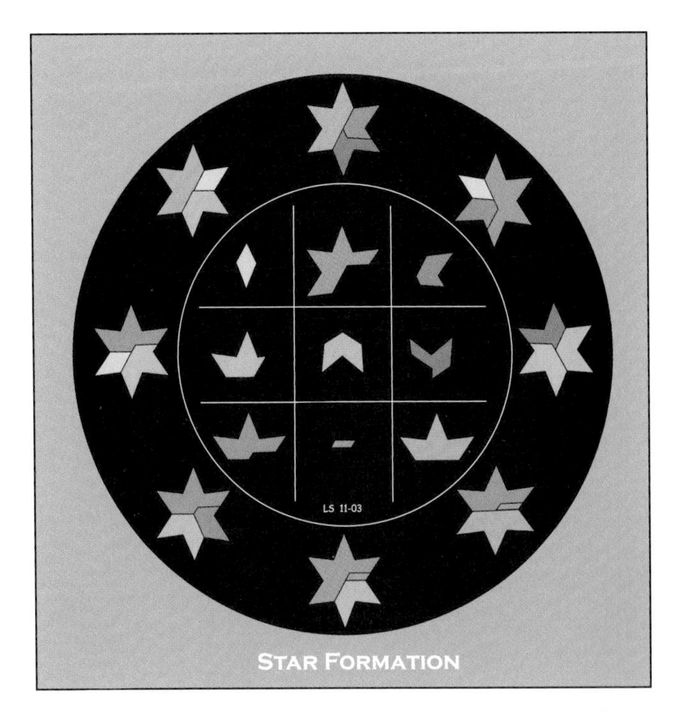

Fig. 4.14 'Star Formation' Evil rats on no star live.

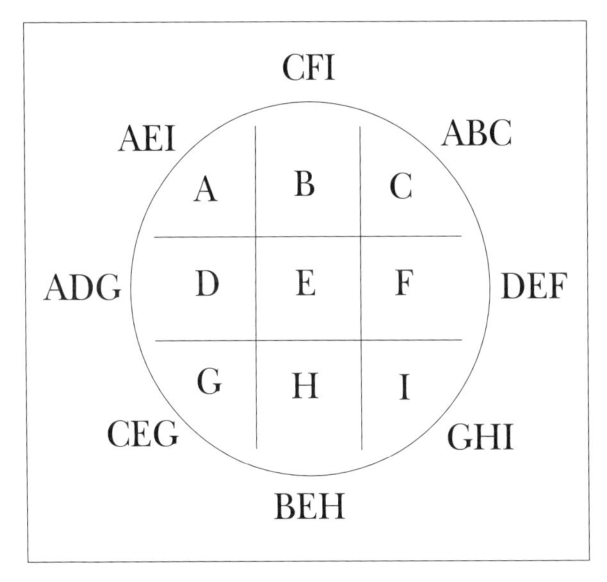

Fig. 4.15 Key to 'Star Formation'.

Consider the 3×3 square in Figure 4.16. The keys and keyholes belonging to each of the three pieces occupying the main diagonal (\) re-echo the congruent piece profiles of Figure 4.3, except in this case applied to circular segments rather than rectangular pieces. That is, the projecting profile on one radial edge is the image of the indented profile on the other radial edge. It is informative to reconstruct the algebraic template of which this square is a geometrical interpretation. To this end, substituting distinct variables *A*, *B*, and *C* for the three different segment sizes (60°, 180°, 120°, respectively), with a and b for the two distinct keys, brings to light the algebraic square of Figure 4.17. It is a square that will not be magic unless the sum of the entries occupying the co-diagonal (/) is made equal to the sum of the three entries in every other line. That is, when $3C = A + B + C$, or $C = \frac{(A + B)}{2}$.

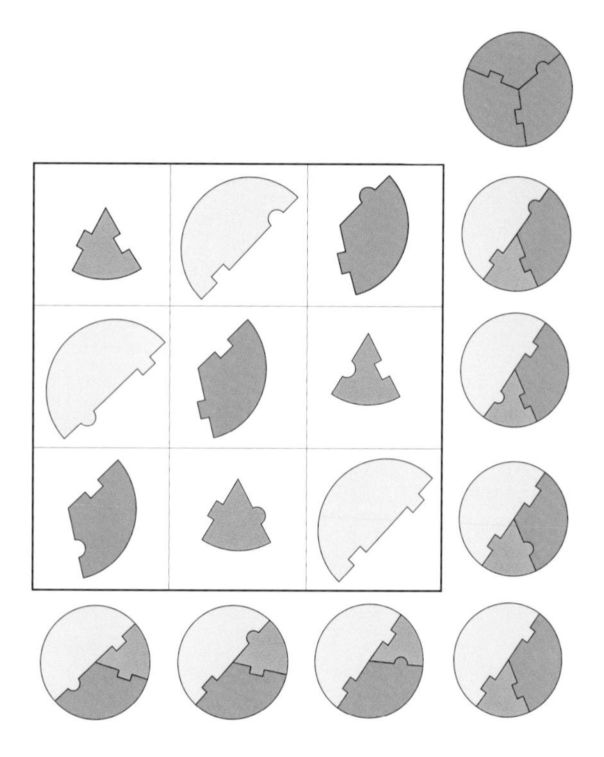

Fig. 4.16 A square using circular segments of three sizes.

$A+a-a$	$B-a-b$	$C+a+b$
$B+a+b$	$C+a-a$	$A-a-b$
$C-a-b$	$A+a+b$	$B+a-a$

$$C = (A+B)/2$$

Fig. 4.17 A template for Figure 4.16.

The result is then a trivial square formed by A, B, and $(A+B)/2$ accompanied by a detrivializing pattern of a's and b's that includes a curious feature. The three entries on the main diagonal all contain '$a - a$ ', a term we would normally ignore or omit because redundant, but is here an intrinsic and necessary part of the template. Such a square reminds us that in applying the template technique we wander somewhat from the path of everyday mathematics and enter a weird world in which the very *mode of expression* used to identify relations becomes as important as those relations themselves. For example, as the reader may like to verify, Figure 4.17 can be shown to be simply an alternative expression of Lucas's formula, which is to say, a square of type 0. But whereas it supplies a template for the 2-D square in Figure 4.16, Lucas's formula emphatically does not, even though the two algebraic squares are mathematically isomorphic.

Little wonder then that the devising of algebraic templates is something of an art, involving at times an uncomfortable reliance on intuition tempered only by trial and error. This is a curious development in what was supposedly to be an exercise in algebra. In defence, I can only say that the challenge confronted in creating geomagic squares has proved too demanding in every other direction, save that of brute force searches using a computer. But better, I thought, a slipshod method that produces results of some kind, rather than a more respectable approach that yields none. And whatever its shortcomings, the template method has certainly proved itself fruitful, as I hope the many examples to be found in these pages will attest.

The brute force searches by computer just referred to are applicable only in the case of squares using polyforms. Special instances of the latter are polyominoes, polyiamonds and polyhexes, which produce tilings showing a greater regularity than that found in other cases. It is this regularity that enables computer programs to be written that are able to identify all the tilings of a given target using a given set of pieces. However, less regular polyforms exist, as exampled in Figure 4.18, which is again based on the template of Figure 4.17.

Here the target is a 'medallion' due to Michael Hirschhorn, who discovered it in connection with some teaching work involving pentagonal tiles conducted at the University of New South Wales in Australia in 1976[6] [7]. The medallion is really the central hub of an infinite tessellation showing six-fold rotational symmetry; it can be extended outwards indefinitely so as to cover the entire plane. The basic tile used is an equilateral pentagon that is among ten pentagonal tiles independently identified at around the same time by Marjorie Rice, a San Diego housewife with a mathematically inventive bent [8] [9]. Complicated as it may seem at first sight, examination will show how the radial profiles of each piece are exactly mirrored by those in the square of Figure 4.16. Figure 20.13 is yet another example modelled on the same template.

There is, in fact, a good deal more to be said about the construction of geomagics based on the template technique. However, the peculiarities of order-3 squares make them an unsuitable vehicle for explaining certain points. In the section on 4×4 geomagics we return to this topic, but for now we pass on.

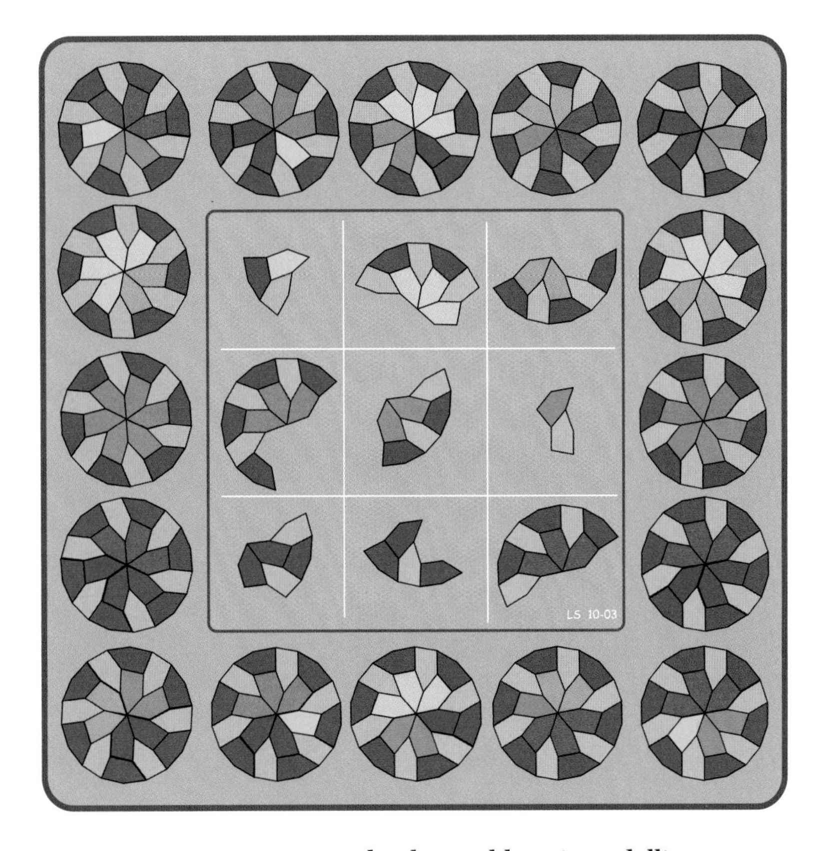

Fig. 4.18 A square using Michael Hirschhorn's medallion as target.

5 Construction by Computer

The five area types discussed before are critical in designing a program able to seek for 3×3 polymagic squares, since besides the need to specify the target shape required, we must also specify the sizes of the pieces to be used. An example will clarify how one such program works.

Suppose we seek a 3×3 polymagic square P using nine polyominoes of equal size; i.e. the area square is of type 4 above. From Lucas's formula we know that the area of its target is an integer divisible by 3, making a 3×5 rectangle, say, a suitable choice of shape. All pieces will then be of size $15 \div 3 = 5$, or pentominoes, of which there exist exactly 12 distinct exemplars. Number these from 1 to 12. A brute force search for a set of 9 pentominoes able to construct P is then simple enough, given first a list L of all the possible sets of 3 distinct pentominoes that will tile a 3×5 rectangle. Every entry in L is thus a triad of distinct numbers in the range 1 to 12. Taking now every possible combination of three distinct entries one at a time, say $\{R,S,T\}$, $\{U,V,W\}$ and $\{X,Y,Z\}$, we imagine these entered into a 3×3 array to produce a square whose rows are now magic because they contain pentomino triads taken from L that therefore tile the target; see Figure 5.1.

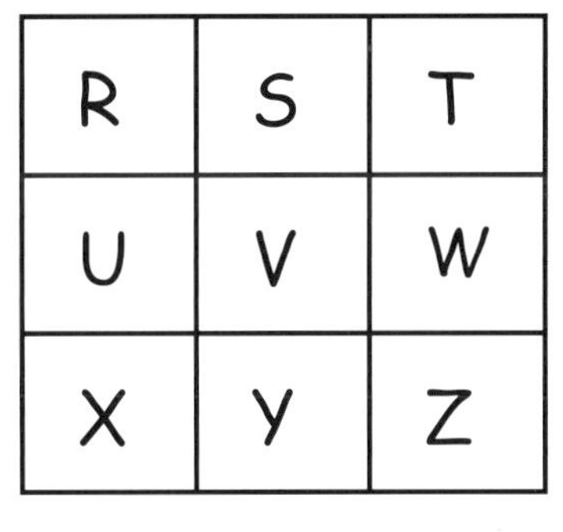

Fig. 5.1 Every row contains a target-tiling triad.

A second routine can now scan L to discover in turn whether the triads $\{R,U,X\}$, $\{S,V,Y\}$ and $\{T,W,Z\}$ are perhaps also on the list. If so, we have found an *ortho*magic square, or one magic on rows and columns only. Following this the diagonal triads $\{R,V,Z\}$ and $\{T,V,X\}$ can be checked and, if it is fully magic, the array saved as a solution. If not, an alternative ordering of the same elements may yet yield a solution. Since the rows are already magic, leaving $\{R,S,T\}$ unchanged, we re-test the array under each of the 6×6 row permutations of $\{U,V,W\}$ and $\{X,Y,Z\}$ in turn, to exhaust all possibilities. With these 36 tests performed, a new combination of three triads from L can be called, say

{*RST*}, {*UVW*} and {*ABC*}, and the process reiterated. Such a program, in fact, reveals there exist no geomagic squares with a 3×5 rectangular target using nine pentominoes; 3×3 polymagics using nine *hexominoes* do exist, as we saw in Figure 2.5 for example. Note that the program just described searches for squares using pieces of identical size. Programs for non-uniform area squares are similar in principle but generally more complicated.

Of course, the difficult part in the above scheme is writing a program to generate the list *L*. In this, I can hardly overstate my indebtedness to Pat Hamlyn, professional programmer and leading name in the field of polyforms, without whose ever generous help my explorations would have been seriously curtailed. At the time, I didn't let on to Pat exactly what I was up to for fear of opening up the topic prematurely. And especially so, in view of the kind of bright polyform-oriented mathematicians Pat hangs out with. If you've been lucky enough to stumble across a previously unknown gold-field then it is only natural to gather up a few of the larger nuggets before rushing into town to let your friends in on the find. So I guess Pat simply assumed I was trying to construct some mechanical puzzles, which would explain my requests for programs that could tell me all the ways that a certain shape could be tiled by certain other shapes. But whatever he thought, he not only provided me with an entire suite of his sophisticated programs, in email after email, time and trouble were not spared in responding at length to my endless questions, and even in adopting his software to my specific demands. Thus, if there is any credit due for tracking down the polymagic squares to be seen in these pages then a very big chunk of it belongs to Pat Hamlyn, the real brains behind the research, and a warm and generous man besides.

So much for a brief sketch of the two present known methods for producing 2-*D* magic squares. We shall now take a look at some example squares of order 3.

6 3×3 Squares

When I first started thinking about how to produce a geomagic square, not knowing any better, I began with the *Lo shu* and tried to create a geometrical analog using polyominoes. The *Lo shu* is a so-called 'normal' square, meaning one using the numbers 1, 2, . . . , N^2, where *N* is the order of the square. Working by trial and error, at length I landed on a solution; see Figure 6.1. The three colors used are merely to assist the eye in distinguishing pieces.

It is an example of what I now term a normal 2-*D* square because the areas of the pieces form the arithmetic progression 1, 2,.., 9, a rare and wonderful property, as it seemed to me at the time. Later on I wrote a computer program able to search for every geomagic square using

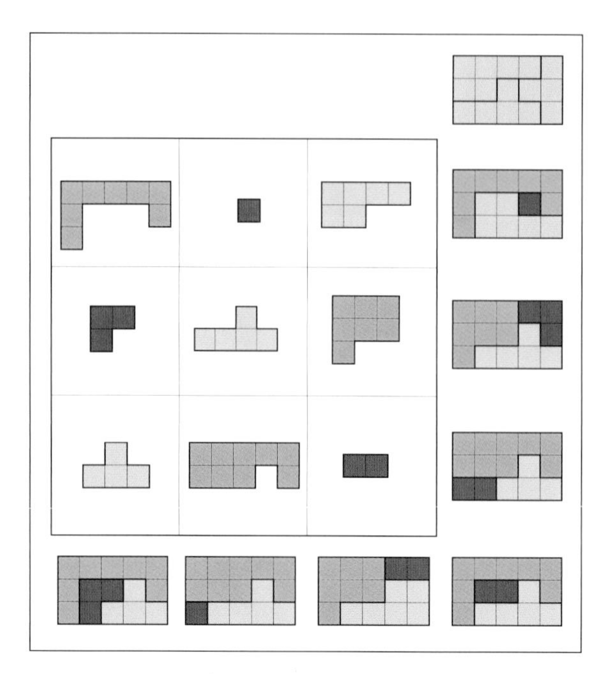

Fig. 6.1 One of 1,411 normal squares with 3×5 rectangular target.

polyominoes with these sizes and same 3×5 rectangular target. The result came as a shock. The above is one of 1,411 distinct solutions, the 8 rotations and reflections of each square not counted. If the target is changed to a 4×4 square missing one of its four centre cells (see Figure 6.2), the computer finds 4,370 solutions.

If the target is a 4×4 square minus one of its inner edge cells (see Figure 6.3) there are 16,465 solutions.

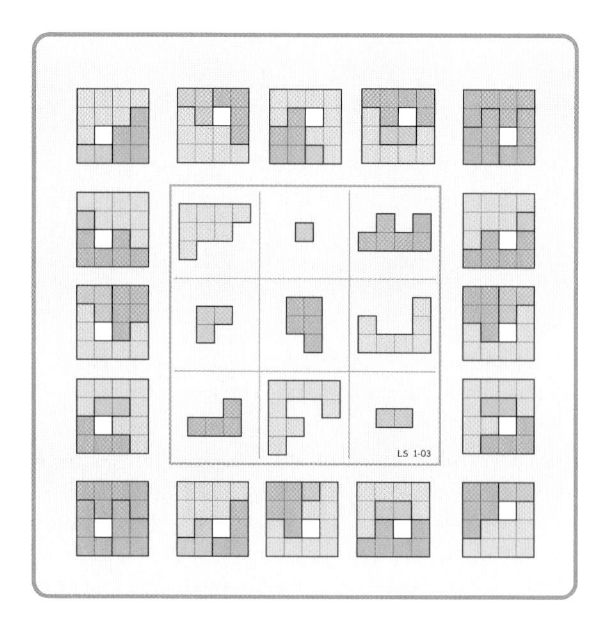

Fig. 6.2 One among 4,370 normal squares using a 4×4 target with inner hole.

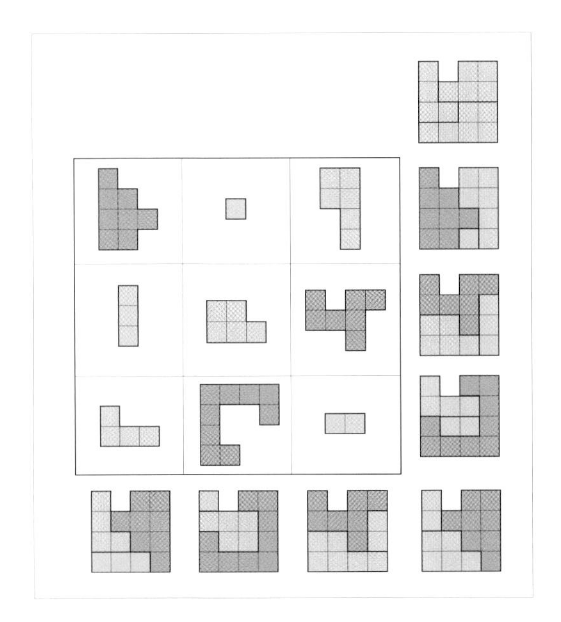

Fig. 6.3 16,465 normal squares share this 4×4 target minus edge cell.

When the target is a 4×4 square minus one of its corner cells then we find 27,110 solutions. Surprising perhaps, but true.

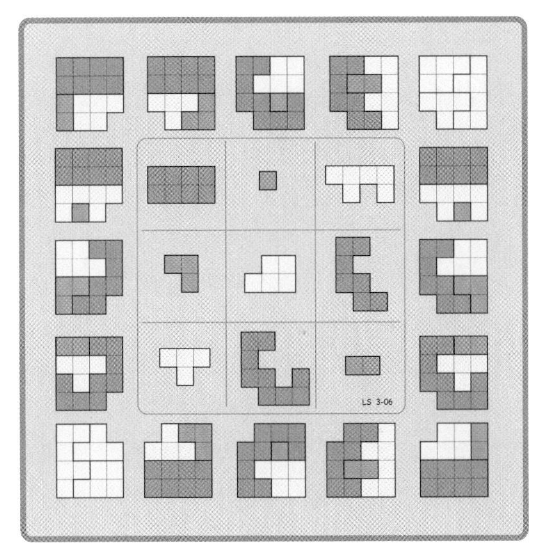

Fig. 6.4 One of the 27,110 normal squares with 4×4 target minus corner cell.

These large numbers of normal squares discovered by computer searches were stored in the form of lists, the items on each list being a set of nine integers that identified the polyominoes appearing in each square. Still later, I wrote another program that could draw these squares one after the other on screen, so as to browse the entire collection and pause on visually interesting specimens. This suggested the idea that a square belonging to the list of solutions associated with one target might also be present in the list for a different target. Or in other words, that a given square might have two distinct targets. A search of the above lists proved fruitless however, although at length I did succeed in finding such a 'bi-magic' square; see Figure 6.5.

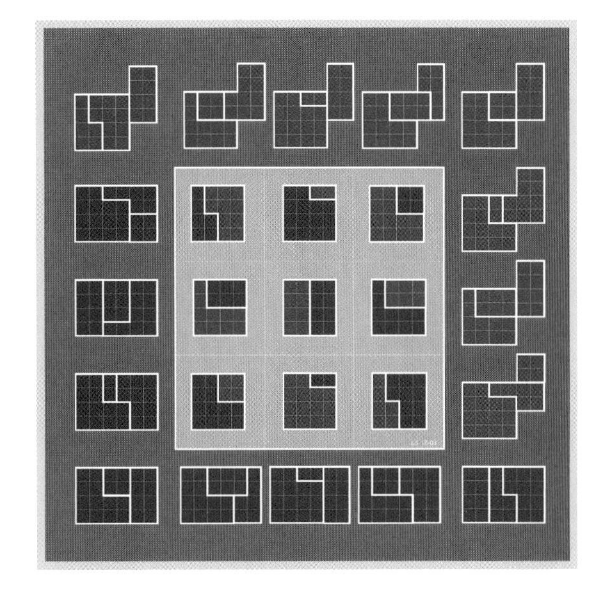

Fig. 6.5 A geomagic square able to tile two distinct targets.

Had I but known it, the pride felt in this seemingly remarkable discovery could hardly have been less appropriate. This would come to light only later, in the wake of a competion held by *Pythagoras*, a well-known mathematics periodical in the Netherlands. *Pythagoras* had published some examples of geomagic squares[10], following which readers were invited to submit geomagic creations of their own. Among those received was the fruit of a combined effort involving three people in three different countries: Odette de Meulenmeester in Belgium, Aad N. J. Thoen from Amsterdam in the Netherlands, and Helmut Postl in Austria. In these days of email, such collaborations are only too easy.

Figure 6.6 shows their amazing discovery. It is a normal 3x3 square using polyominoes, being, in fact, one of the 1,411 solutions with 3×5 rectangular target mentioned above. But as Figure 6.6 clearly demonstrates, this rectangle is far from being the only target that can be assembled with these pieces. Incredibly, it is one of *twelve* alternative shapes that may be used! Little wonder that it took three collaborators to identify this gem: Thoen to discover an initial five targets, de Meulenmeester to find three more, followed by Postl with the final four. I can only admit to astonishment and offer my congratulations.

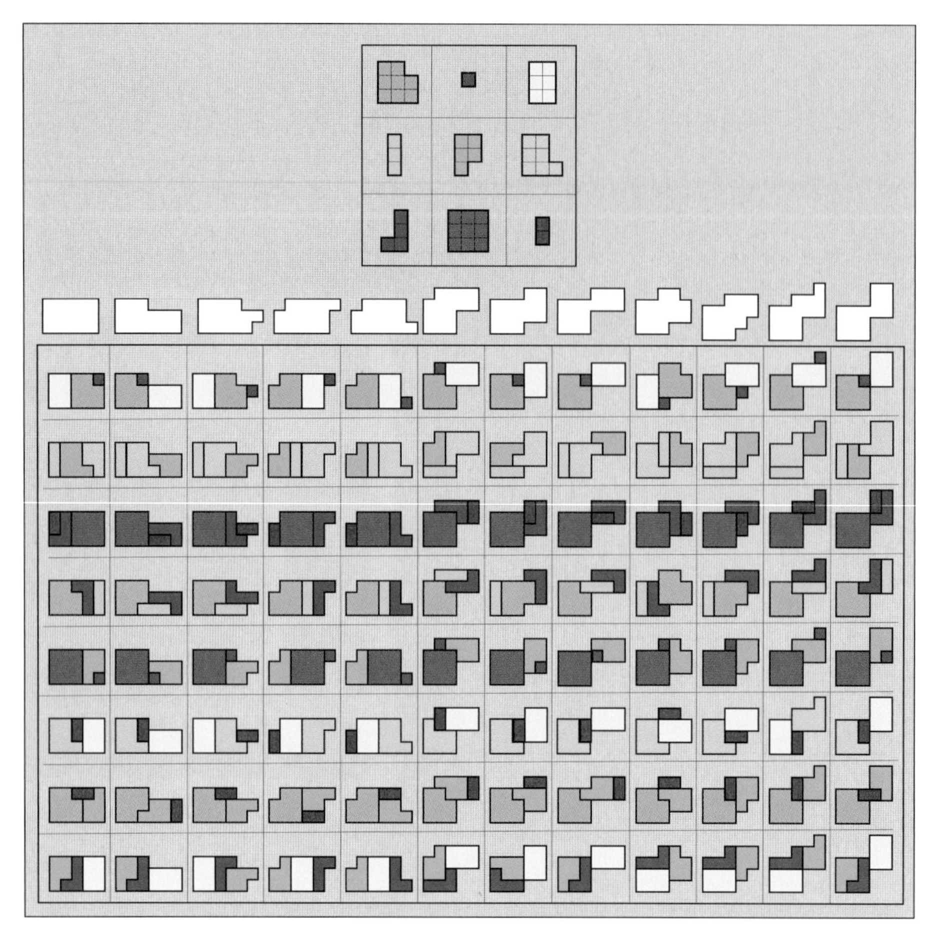

Fig. 6.6 A normal square able to tile twelve distinct targets.

Note an important difference between Figures 6.1–4 and the geomagic square with which we started, Figure 2.2. Lucas's formula shows us that every 3×3 *numagic* square is a so-called 'symmetric' or 'associated' square. That is, the sum of any two numbers diametrically opposed about the centre is $2c$, or twice the centre number. Remarkably, Figure 2.2 is *geometrically* symmetric: each of the 4 pairs of diametrically opposed pieces will together tile a rectangle of 4×6, which is also tiled by two copies of the 2×6 centre rectangle. Figure 6.7, an alternative presentation of Figure 2.2, seeks to highlight this property, which is absent from Figures 6.1–4, as it is from most 3×3 geomagics. Figures 4,10, 4.11 and 6.5 however yield three further examples of symmetric squares. In the latter the two complementary pieces will combine to complete a 4×4 square.

Readers who, like me, have a preference for square over rectangular targets might surmise that the building blocks of polymagic squares could be deformed so as to achieve this result whenever required. Consider Figure 6.1, for example. Squashing the 1×1 unit squares forming each piece into unit rectangles of width 3/5 ought to do the trick, for then the shapes must contract horizontally

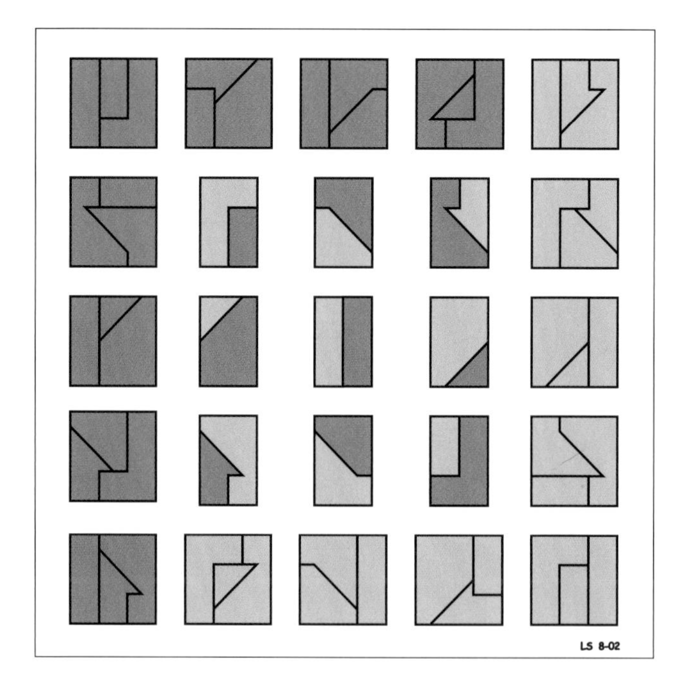

Fig. 6.7 Complementary piece pairs tile the same rectangle.

so as to yield the desired square target. However, this will work only so long as each piece occurs in the *same orientation* within every target. But a look at the tetromino in the bottom left-hand cell of Figure 6.1, say, shows that here this is not the case. In the bottom row and left-hand column targets it appears unchanged in orientation, but in the co-diagonal (/) target it is rotated. Following horizontal squashing, it would therefore be *too short* to completely tile the latter target.

In fact, experience teaches that the interlocking relations within geomagics can be very deceptive. Again and again, one feels convinced that some insignificant detail can be altered, only to find out that the change has disastrous consequences elsewhere in the square. Figure 6.8(a) is one among a few examples found of a geomagic square in which the pieces appear neither rotated nor reflected in any target. This is perhaps unsurprising; just as a conventional magic square may contain the number zero, so Figure 6.8(a) makes use of the 'empty' piece, with the result that only eight pieces are involved. The constant orientation of these eight pieces means that the entire square can be stretched or squashed without affecting its geomagic properties, as shown in Figures 6.8(b) and (c). The target can then be a square (b), or even a parallelogram (d).

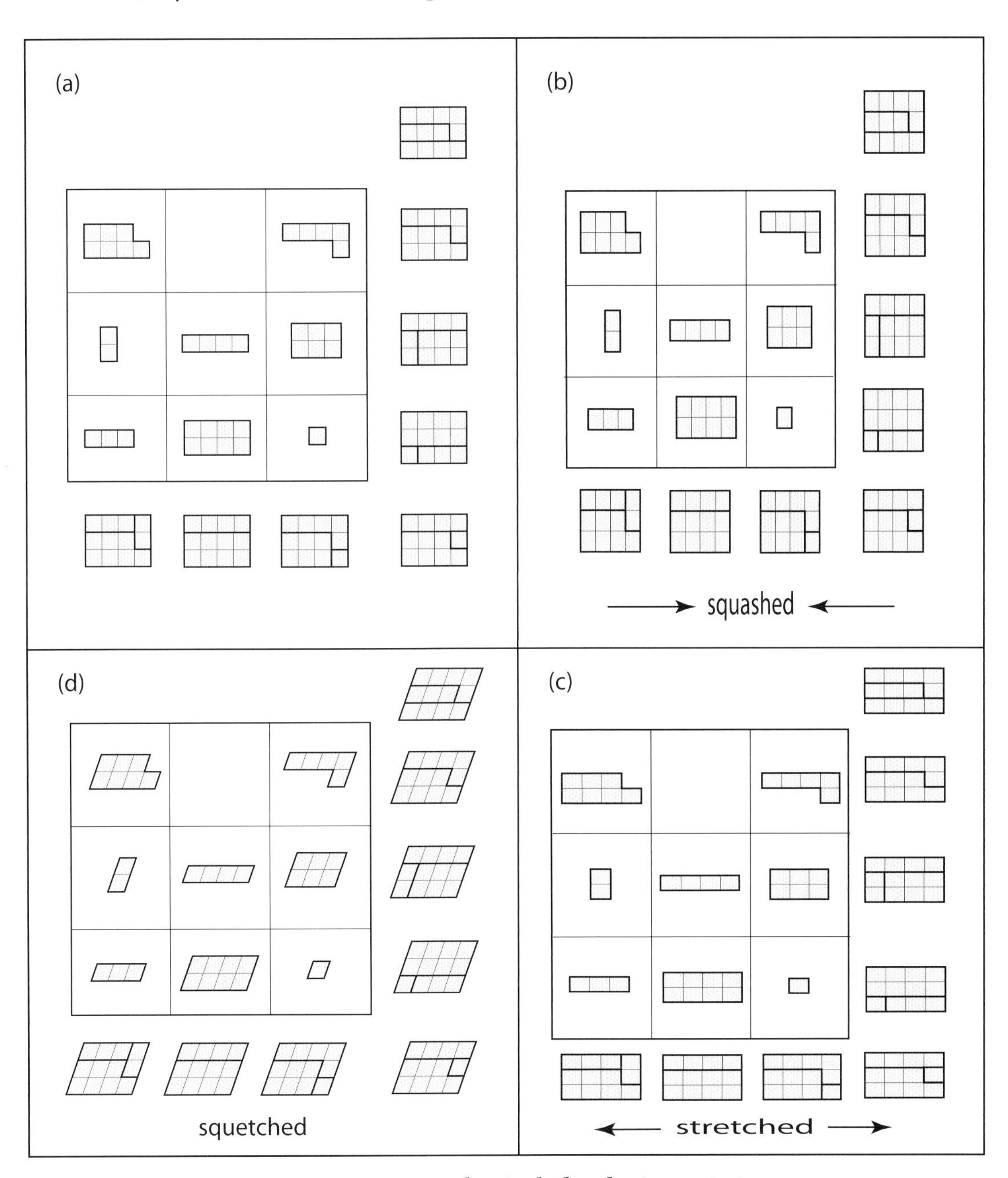

Fig. 6.8 A square that includes the 'empty' piece.

7 3×3 Nasiks and Semi-Nasiks

Often geomagics with some desired special property can be found only at the price of accepting less pleasing targets. Consider for instance Figure 7.1, in which the region tiled is a 4×5 rectangle missing two edge cells.

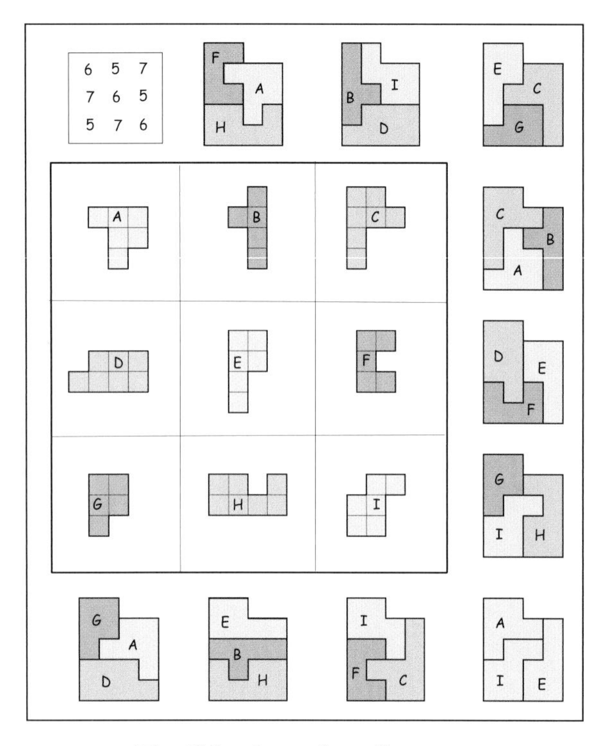

Fig. 7.1 A semi-nasik square.

The pieces used are of three sizes: 3 pentominoes, 3 hexominoes, and 3 heptominoes, the areas forming a *Latin square*, seen above left. The subject of Latin squares has a long history, abounding as it does with unsolved problems, some of them as many as 200 years old [11]. By a Latin square of order N, we refer to a square of N^2 entries composed of N distinct elements, each of which occurs exactly once in every row and column. Here the entries happen to be numbers, but could equally be elements of a different kind, such as letters or geometrical shapes. We shall have more to do with Latin squares when we come to 4×4 geomagics.

Figure 7.1 is an example of a square that is 'semi-panmagic' or 'semi-nasik'. Fully panmagic or nasik[2] squares (so named after the town in India where an early 4×4 specimen was found) are those in which every diagonal, including the so-called 'broken' diagonals, *AFH*, *BDI*, *CDH*, and *BFG*, are also magic lines, which is to say, their pieces tile the target. Note that *AFH* and *BDI* are parallel and hence non-intersecting, as are *CDH* and *BFG*. Semi-nasik squares of 3×3 are those showing a total of 4 magic diagonals that include the two main diagonals plus one of these two parallel pairs. In Figure 7.1 the latter are *AFH* and *BDI*, as shown by the targets at top.

There exist similar 3×3 geomagics with a total of 4 magic diagonals that are different to those in semi-nasiks. These are squares in which the two main diagonals, plus two broken diagonals that are non-parallel and thus intersecting are magic. I call these *demi*-panmagic or *demi*-nasik squares. Figure 7.2 shows an example that includes *weakly-connected* pentominoes, or those in which some unit squares join only at their corners. On referring to Lucas's formula, we see that a magic diagonal such as *AFH* will imply $(c + a) + (c + a - b) + (c + a + b) = 3c$, from which is found $b = 0$. Continuing in the same way, it is simple to verify that semi-nasiks must have area squares

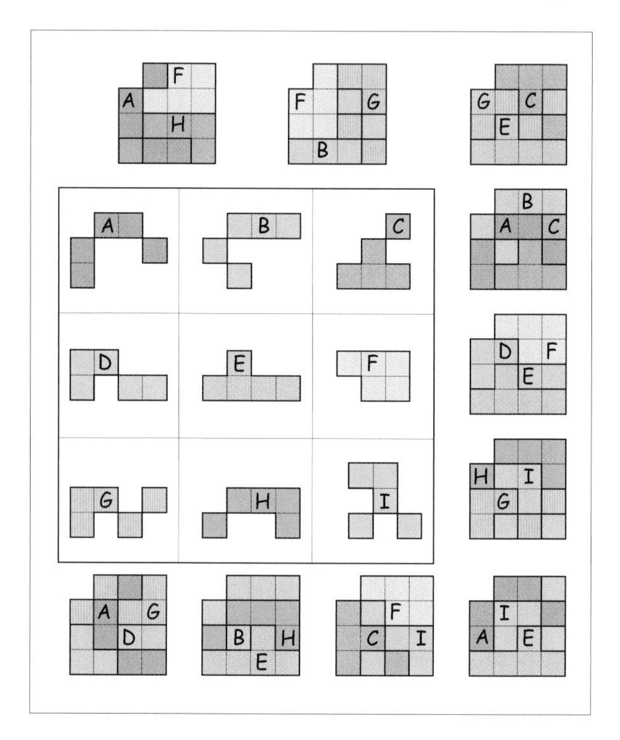

Fig. 7.2 A demi-nasik square.

of type 3 or 4, while those of demi-nasiks can be of type 4 only.

The semi-nasik property of Figure 7.1 is remarkable for the reason that, as is well known (and again, easy to ascertain with the help of Lucas's formula), there exist no non-trivial 3×3 nasik or semi-nasik *numerical* magic squares[3], although such can be found for all higher orders. Figure 7.1 is thus a potent demonstration that geomagics listen to laws different to those governing ordinary magic squares. Alas, attempts to find a 3 × 3 semi-nasik square

2 Nasik squares are also known as pandiagonal, diabolic, or satanic.

3 Whereas semi-nasik 3 × 3 multiplicative numagic squares do exist (using complex cube roots of unity), a fact nowhere previously recorded in the literature, so far as I am aware.

sporting a more symmetrical target have thus far come to nothing.

The discovery of semi-nasiks led inevitably to a search for fully nasik squares, a quest that was eventually successful, but only at the price of introducing *disconnected* polyominoes. Among the 5 area types, only type 4, showing same entries in every cell, enjoys 6 magic diagonals. In any nasik geomagic square, pieces will thus be of uniform area. Figure 7.3 shows an example including both weakly-connected and fully disconnected pentominoes. This square is remarkable for a further reason. Choose any three pieces belonging to any three of the four corner cells. There are 4 possible choices. The selected triad will tile the target in every case. The resulting 16 near-square targets that encircle the 3×3 array make for a pleasing mathematical ornament. A key to the square is shown in Figure 7.4.

It is a well-known property of nasik squares of any size that they remain nasik under cyclic permutation of their rows and columns. This is nicely demonstrated when the plane is tiled with repeated copies of an $N \times N$ nasik square, with the result that *any* arbitrarily selected $N \times N$ area will again be found to be a nasik square. This is illustrated in Figure 7.5 using the above 3×3 specimen.

The identification of 3×3 nasik and semi-nasik geomagics —there are many to be found beside those shown here, as well as squares showing 0, 1, 3, or 5 magic diagonals— raises an interesting question. As seen with the nasik-preserving cyclic permutations of rows and columns, unlike standard 3×3 geomagics, the entries in nasik and semi-nasik squares can be rearranged to yield still more geomagic squares. But exactly how many distinct specimens can be thus formed? The non-existence of numagic nasiks or semi-nasiks means that this question has never before been addressed. A simple computer program provided the answer by examining in turn every permutation of the 9 letters in Figure 7.4 (or Figure 3.1), which is interpreted as representing a nasik square exhibiting the 12 magic triads: *ABC, DEF, GHI, ADG, BEH, CFI, AEI, BFG, CDH, CEG, AFH* and *BDI*.

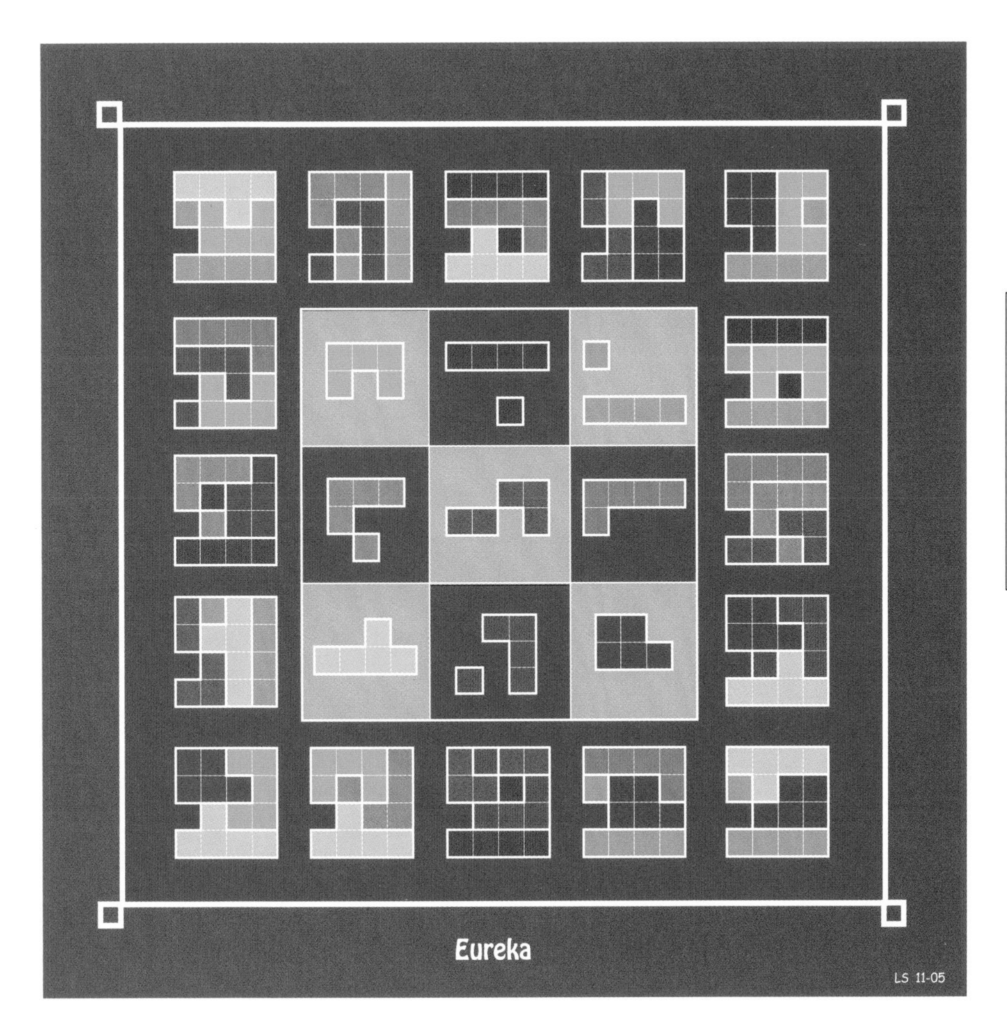

Fig. 7.3 A nasik square of order-3.

ACG	CDH	BFG	AEI	ACI
AFH	A	B	C	ABC
BDI	D	E	F	DEF
CEG	G	H	I	GHI
AGI	ADG	BEH	CFI	CGI

Fig. 7.4 Key to Fig 7.3

Fig. 7.5 Tiling the plane with copies of a nasik square means that any randomly selected 3×3 area is again a nasik square. Three examples are outlined.

Nasik-preserving permutations of the letters will be those in which every row, column and diagonal is again occupied by one of the listed 12 triads. A similar method established the number of semi- and demi-nasik rearrangements. In comparing the squares identified by the program, it took but a little detective work to unravel their relations so as to describe these in terms of a few basic transformations. As a matter of fact, these investigations were performed before ever a nasik or semi-nasik geomagic square had actually been discovered. I was thus in the curious position of having a pretty thorough understanding of 3×3 nasik and semi-nasik geomagic squares long before even knowing whether or not any existed.

The number of magic rearrangements for semi-nasiks is nine, rotations and reflections not counted. Two transformations (*T*1, *T*2), both of order-3, generate all nine, as diagrammed in Figure 7.6(a.) *AFH* and *BDI* are the two magic broken diagonals. Demi-nasiks yield just four squares, as generated by transformation *T*3 in Figure 7.6(b.) Nasik squares are still more prolix, *T*1 and *T*2 combining with the two transformations *T*4 and *T*5 shown in Figure 7.6(c,) as well as two further transformations that produce the 8 rotations and reflections of each square, to form a group of order-432. Thanks are due to my friend Michael Schweitzer for identifying this as the affine general linear group *AGL*(2,3). In consequence, the entries in any 3×3 nasik geomagic square can always be permuted to produce at least $432 \div 8 = 54$ distinct squares, rotations and reflections not counted. Squares that result from permuting pieces in nasik or semi-nasik squares are themselves always nasik or semi-nasik, respectively.

Surprising as it may seem, there are still more magic triads to be found in Figure 7.3 than the 16 appearing in the targets drawn. In all there are 36. They are as follows:

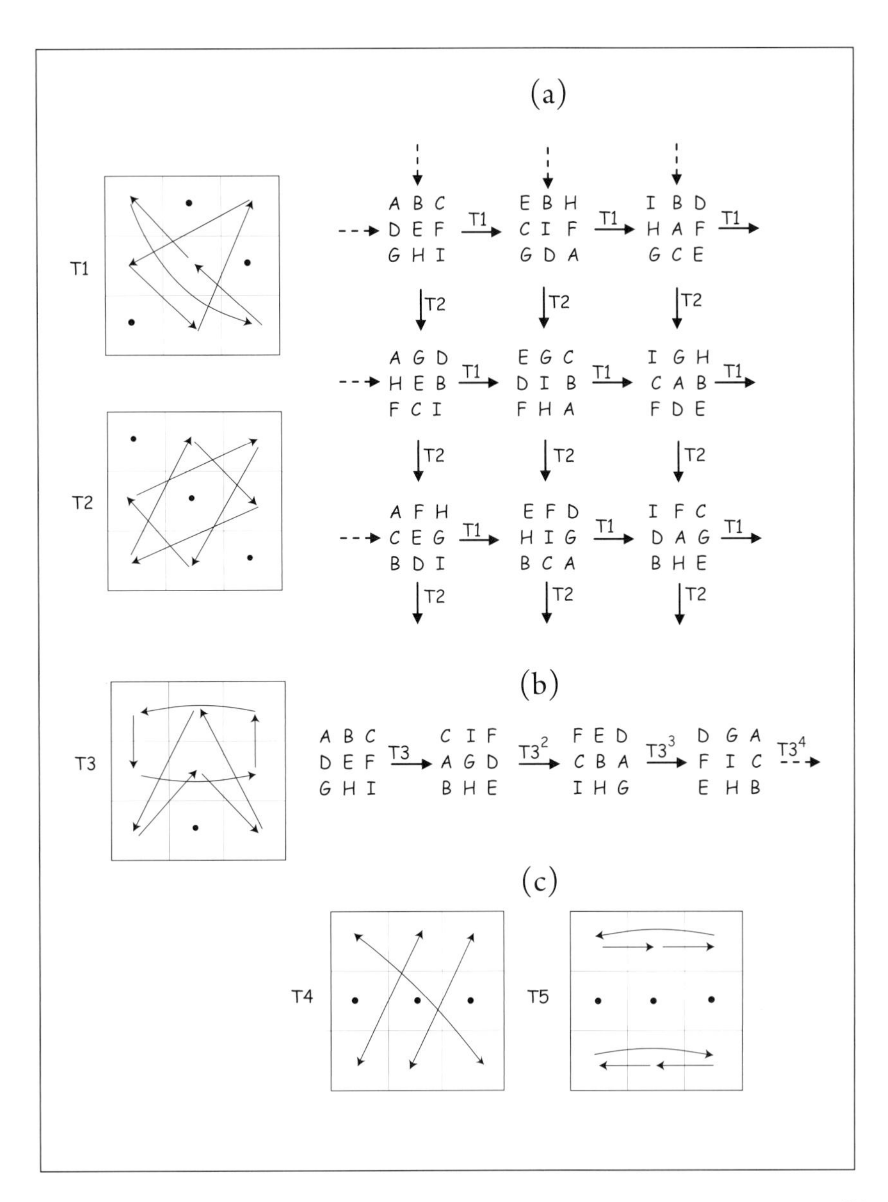

Fig. 7.6 (a) Rearranging the pieces in a semimagic square gives rise to 9 variants. (b) Demi-magic rearrangements number 4. (c) Transforms *T1*, *T2*, *T4*, and *T5* yield 54 variants of a nasik square.

ABC	*ABI*	*ACI*	*AEH*	*AFH*	*AGI*	*BDH*	*BEH*	*BHI*	*CEG*	*DEF*	*EFI*
ABD	*ACG*	*ADE*	*AEI*	*AFI*	*BCE*	*BDI*	*BFG*	*CDE*	*CFI*	*DGH*	*FGI*
ABF	*ACH*	*ADG*	*AFG*	*AGH*	*BCI*	*BEF*	*BFI*	*CDH*	*CGI*	*EFH*	*GHI*

We have seen that the pieces in Figure 7.3 can be rearranged so as to yield at least 54 different nasik squares. Might it be that this large number of extra magic triads will allow even more to be formed? If so, the total must be a multiple of 54, since for every extra square, there will be its accompanying 53 nasik permutations. In fact a computer trial shows that this is not the case with Figure 7.3. But squares having this property can be found. The pieces in Figure 7.7, for example, another nasik square with a smaller number of magic triads, can be rearranged so as to yield 2 × 54 = 108 nasik squares. Its 31 magic triads are as follows:

ABC	*ADF*	*AFG*	*BDI*	*BFH*	*CDH*	*CFI*	*DEF*	*DFI*	*EGH*	*GHI*
ABD	*ADG*	*BCE*	*BEH*	*BGH*	*CEF*	*CFG*	*DEG*	*DGI*	*FGH*	
ABH	*AEI*	*BDE*	*BFG*	*BHI*	*CEG*	*CGI*	*DEH*	*EFH*	*FGI*	

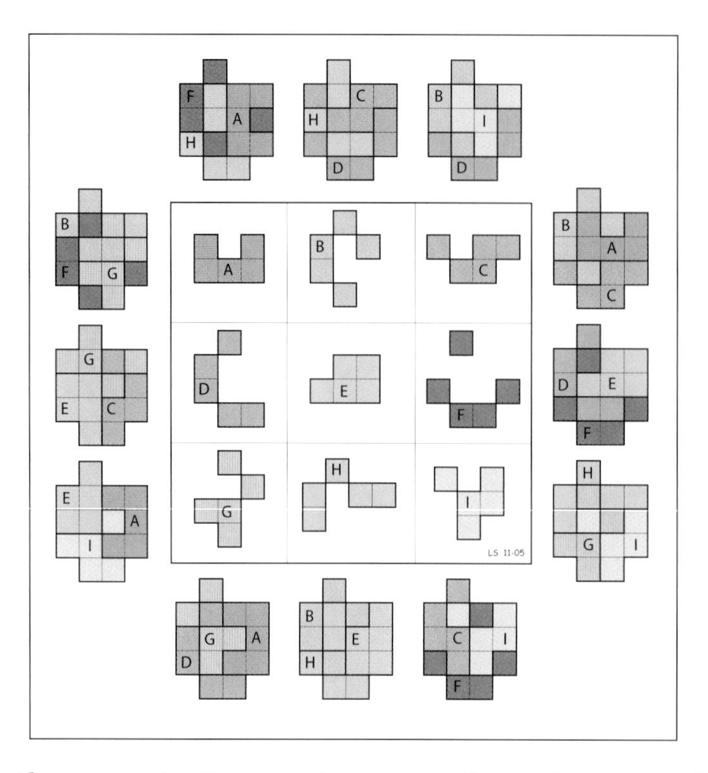

Fig. 7.7 A nasik square using 9 pentominoes, normal, weakly-connected, and disjoint.

The reason for this doubling in numbers is not difficult to find. As the reader can verify, pieces *E* and *I* in Figure 7.7 can be switched, yet the square remains geomagic. Scrutiny shows that this will be possible only when the 6 triads, *DFI*, *BHI*, *CGI*, *CEF*, *EGH*, and *BDE*, are among the 31. This is a consequence of the fact that, following such a switch, all six of these triads will find themselves occupying rows, columns and diagonals. Hence a nasik square might have as few as just six extra magic triads and still yield 108 different squares.

8 Special Examples of 3×3 Squares

Above we noted the existence of thousands of geomagics using pieces with areas of 1, 2, . . . , 9 units. Since the areas are all different, their area squares are of type 0. Research indicates that the fertility[4] of a set of pieces falls dramatically as the number of same-sized pieces in the set increases. For example, the pieces in a type-2 square (page 15), exhibit just five different areas. Figure 8.1 shows an example; the pieces are of sizes 4, 6, 8, 10 and 12. It is one of only three solutions found. 3×3 squares in which all the pieces are of same size are even rarer. Figures 2.5 and 8.5 show examples using nine hexominoes and decominoes, respectively.

4 A detailed discussion of fertility can be found in my article 'New Advances with 4×4 Magic Squares', which is included as Appendix III.

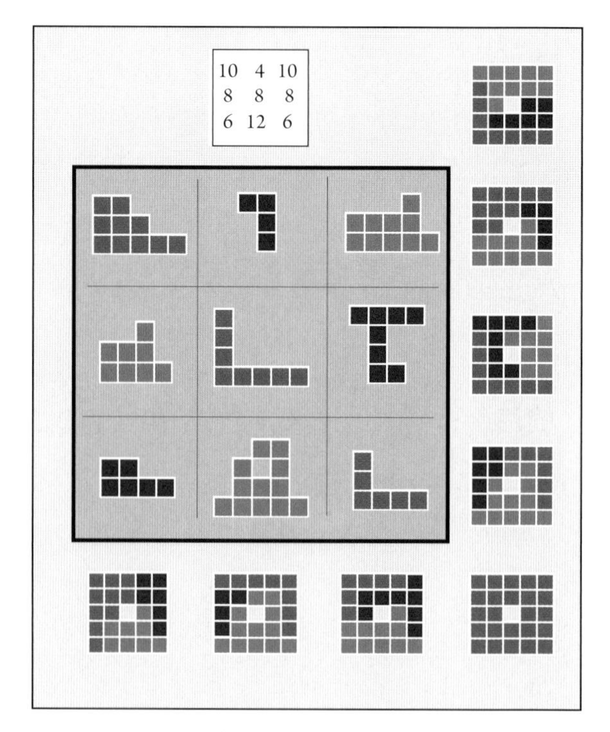

Fig. 8.1 Repeated piece sizes imply fewer solutions.

The target with central hole in Figure 8.1 is less a decorative flourish than a consolation prize. To see why, suppose we seek a 3×3 polymagic square showing a *solid* square target. The target area must then be a square number that is a multiple of 3, or 3 times the area of the centre piece. The possibilities are thus 9, 36, 81, ... However, 9 is impossible because the pieces required would be too small to allow enough distinct shapes. 81 implies an average piece size of 81 ÷ 3 = 27, which is too large for a personal computer to handle because the numbers of piece combinations becomes prohibitively large. 36 seems hopeful. Let us begin with a search for a square using nine dodecominoes, or polyominoes formed of 12 unit squares. The program requires a list L of all the triads of dodecominoes that tile a 6×6 square. There are 32,222 such triads. Being a longish list, it takes the program a while to check whether the triad of pieces in a candidate column/diagonal is, or is not, in L. And with 36 permutations of each candidate square to test, there are a lot of checks to perform. Such a program ran for weeks on my PC, without finding a solution. Figure 8.2 shows one of several *simple* or orthomagic squares discovered along the way. The latter are squares that are magic on rows and columns only.

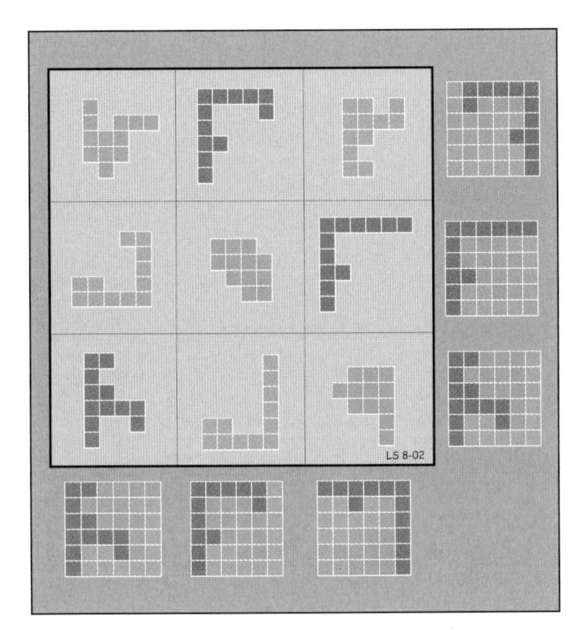

Fig. 8.2 An orthomagic square using same-sized pieces.

A next step might be to search for a square using three 11-ominoes, three 12-ominoes (or dodecominoes), and three 13-ominoes, as in the area square of type 3 shown in Figure 8.3. The area of the target is then 11 + 12 + 13 = 36, as required. Alas, the program that I use to generate lists cannot handle polyominoes larger than 12-ominoes.

There is a good reason for that; the number of resulting piece combinations becomes simply too huge to handle. So you see, the little matter of finding a polymagic square with a square target is no trivial task. After a little thought, the 5×5-square-minus-centre-cell target with area 24 then suggests itself as a next best choice. Even so, searching for such a polymagic square is far from simple. With a target area of 24, the area of the centre piece must be 8. The posssible piece area schemes (using pieces not larger than 12) are then as shown in Figure 8.4.

11	13	12
13	12	11
12	11	13

Fig. 8.3 An area square of type 3.

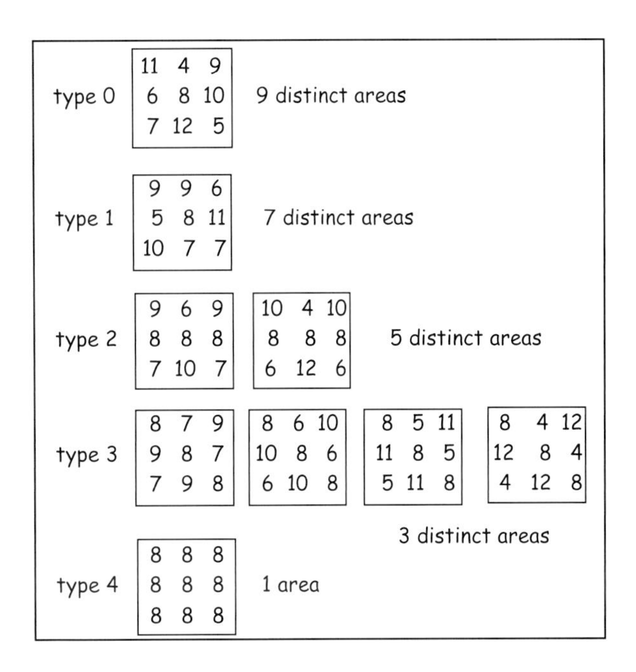

Fig. 8.4 Piece area schemes for a target of area 24.

A dedicated program must examine each of these cases separately. In the program that searched for and found Figure 8.1, there is not one list L needed, but five, corresponding to the 5 different triads found in the rows, columns and diagonals of the area square in Figure 8.1, the second of the type-2 schemes listed in Figure 8.4:

L_1 lists all target-tiling triads using one 4-omino and two 10-ominoes (top row).

L_2 lists all target-tiling triads using three 8-ominoes (centre row),

L_3 lists all target-tiling triads using two 6-ominoes and one 12-omino (bottom row).

L_4 lists all target-tiling triads using one 6-omino, 8-omino, and 10-omino (2 columns + 2 diagonals),

L_5 lists all target-tiling triads using one 4-omino, 8-omino, and 12-omino (centre column).

Testing candidate squares entails checking the columns and main diagonals to see if the three pieces tile the target, which is to say, are present on their corresponding list.

At the extreme of area variability are squares using nine pieces of same area, which are rarest of all. Figure 8.5 shows a polymagic square using nine decominoes; the target is a rectangle of 5×6. It is one of two solutions. Although polymagics using nine smaller equal area pieces can be found (starting with 9 hexominoes), Figure 8.5 is the first I have discovered with a rectangular target.

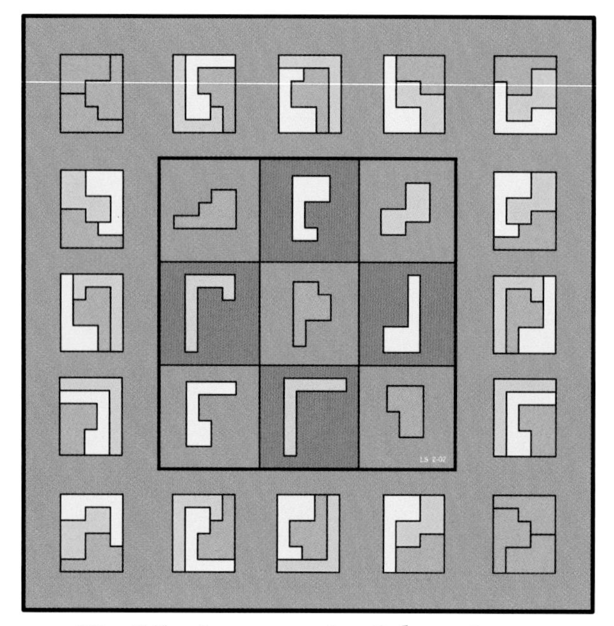

Fig. 8.5 A square using 9 decominoes.

Fig. 8.6 'Diamond Sutra', uses polyiamonds tiling a diamond-shaped target.

The polymagics looked at so far have all used polyominoes. Figure 8.6, using polyiamonds, or shapes constructed from unit equilateral triangles, is among my favourite finds. The target, here drawn at a reduced scale, is an equilateral parallelogram or diamond, while the piece sizes form a consective series, obtained by adding 1 to the entries in the *Lo shu*.

Since the 9 pieces used here form three diamonds, the latter can be put together to form a regular hexagon. Now the area of the diamond is 18 units, and since there are only eight ways in which three integers with a sum of 18 can be chosen from 2, 3, . . . , 10, the targets shown in Figure 8.6 must account for every possible diamond formable using three of these pieces. Three diamonds can be chosen from among the eight possibilities in $\binom{8}{3}$ = 56 ways. Moreover, each diamond can be reflected about either or both of its diagonals so as to yield 4 possible orientations, with the result that any given triad of diamonds can be assembled in 4 × 4 × 4 = 64 different ways so as to complete a distinct hexagon. In total there are thus 56 × 64 = 3,584 distinct hexagons that can be created in this manner. Remarkably, however, the same pieces can be assembled to form a regular hexagon in many other ways, such as the following:

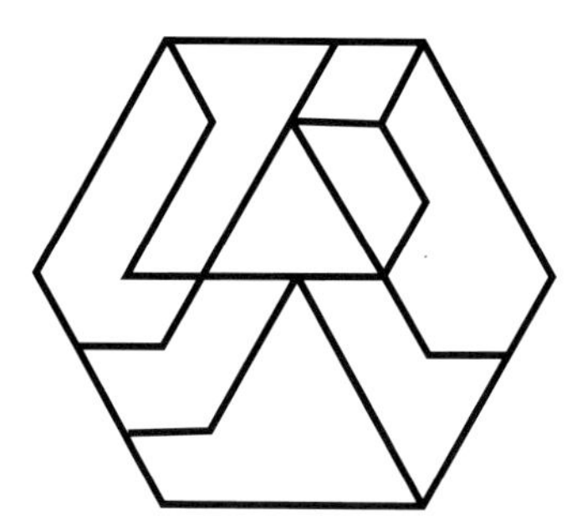

Fig. 8.7 A regular hexagon formed with the Diamond Sutra pieces.

In fact, computer investigation reveals an astonishing 17,213 distinct hexagons that can be constructed in this way, over and above the 3,584 already identified, a total of 20,797 in all.

The title of Figure 8.6, *Diamond Sutra*, is perhaps a bit fanciful, but reflects my romantic view of geomagic squares as objects of contemplation (not to say veneration). The same tendency reappears in the top center cover illustration and Figure 8.8, not merely in their titles, *Magic Mandala I* and *II*, but in their octagonal layout, an idea suggested by a Tibetan astrological diagram at the centre of which a 3×3 magic square is enclosed within a circle surrounded by eight trigrams. The pieces used are again polyiamonds, with targets that are in both cases a regular hexagon. The key in Figure 8.9 identifies the target associated with each row, column, and diagonal. These are the only two solutions using polyiamonds with these sizes and same target shape, the corresponding area squares being in both cases Latin squares. However, using nine pieces of size 4, 5, 6, . . . ,12, there exist 38 solutions in which the target is again a regular hexagon of area 24. It would be difficult to exaggerate my delight in the discovery of these objects. I don't suppose they are of any mathematical significance, but their effect on my aesthetic sense is mesmeric.

I conclude this brief overview of 3×3 geomagic squares with a glance backward to our starting point at Figure 1.1, showing the *Lo shu* in both its conventional and traditional guises. In the latter, lines and dots represent the numbers from one to nine, a device suggestive of great antiquity, but in reality the work of a medieval ironist [2]. In fact, by the time of its composition, it had long been common practice in China to represent numbers by means of characters. It is amusing to see how close this prankster came to unwittingly inventing geomagic squares, as the modern specimen shown in Figure 8.10, borrowing heavily on his own idea of using geometrical patterns to represent numbers, will attest.

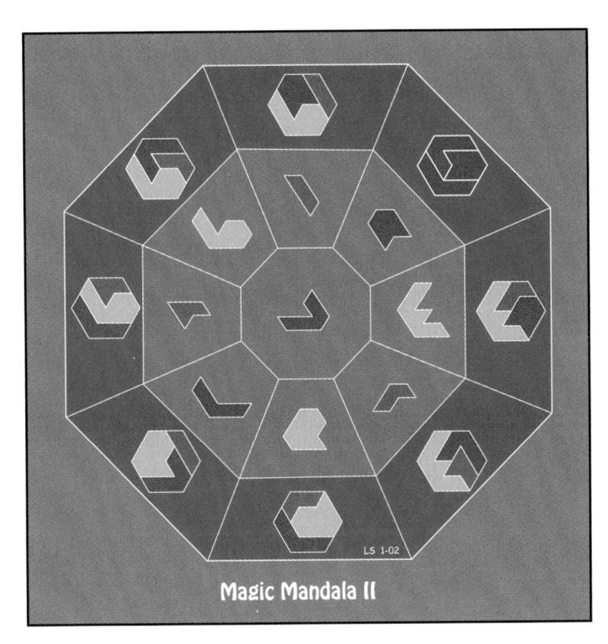

Fig. 8.8 A 3×3 square with hexagonal target and octagonal layout.

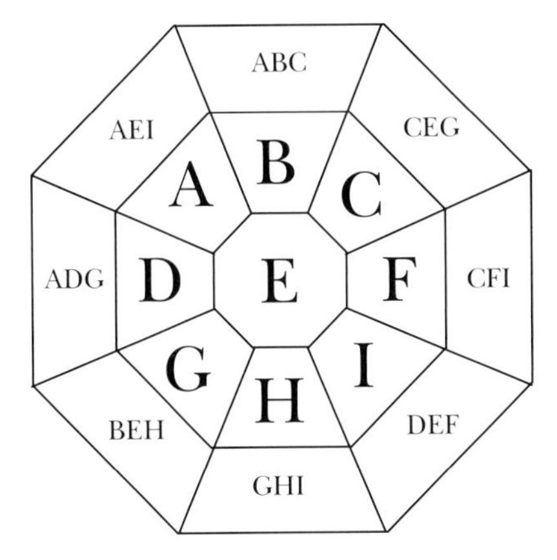

Fig. 8.9 Key to 'Magic Mandala II'.

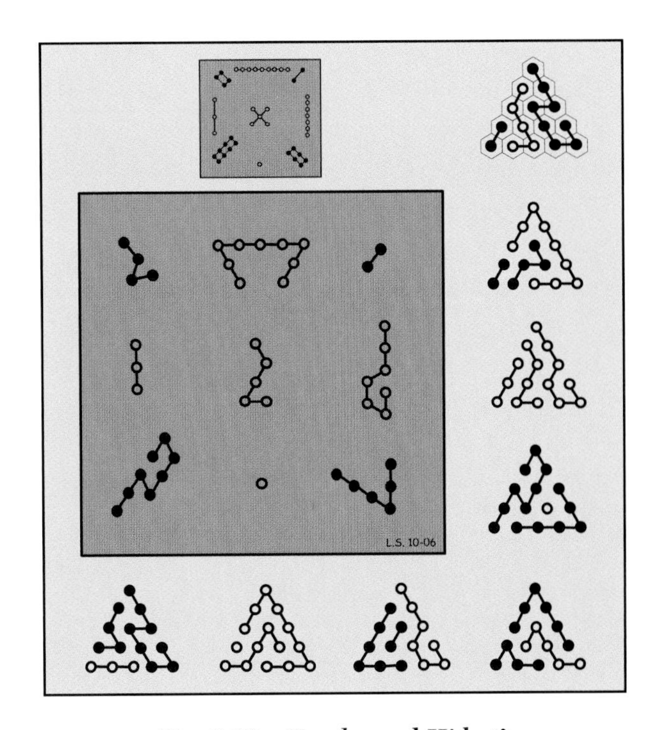

Fig. 8.10 '*Lo shu* and Hi hat'.

It may not be immediately apparent that the pieces appearing in Figure 8.9 are essentially *polyhexes*, or figures composed of unit hexagons. The underlying target, a symmetrical shape of area 15, is thus better interpreted as the triangular honeycomb of Figure 8.11, in which the hexagons might equally be replaced by kissing circles.

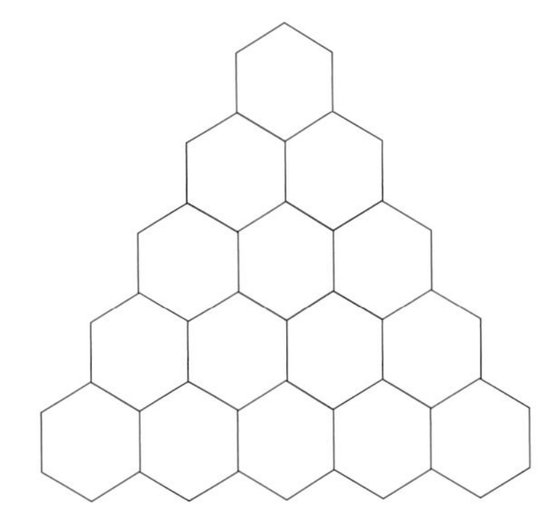

Fig. 8.11 Close-up of the target used in Figure 8.10.

Almost incredibly, Figure 8.10 is one of 169,344 distinct solutions, their rotations and reflections as usual not included, using polyhexes of size 1, 2 , . . . , 9 to produce the same triangular target. If we think of the target outline as a rough approximation to an elongated Chinese peasant's straw sun-hat, then a good title for Figure 8.10 might be '*Lo shu* and Hi hat.'

The lines and dots representation need not be confined to hexagonal grids. Figure 8.12 shows "Chinese Abacus," an alternative version based on polyominoes. Along with Figure 6.1, it is one of the set of 1,411 consecutive area squares with 3×5 rectangular target mentioned earlier (see Figure. 6.1).

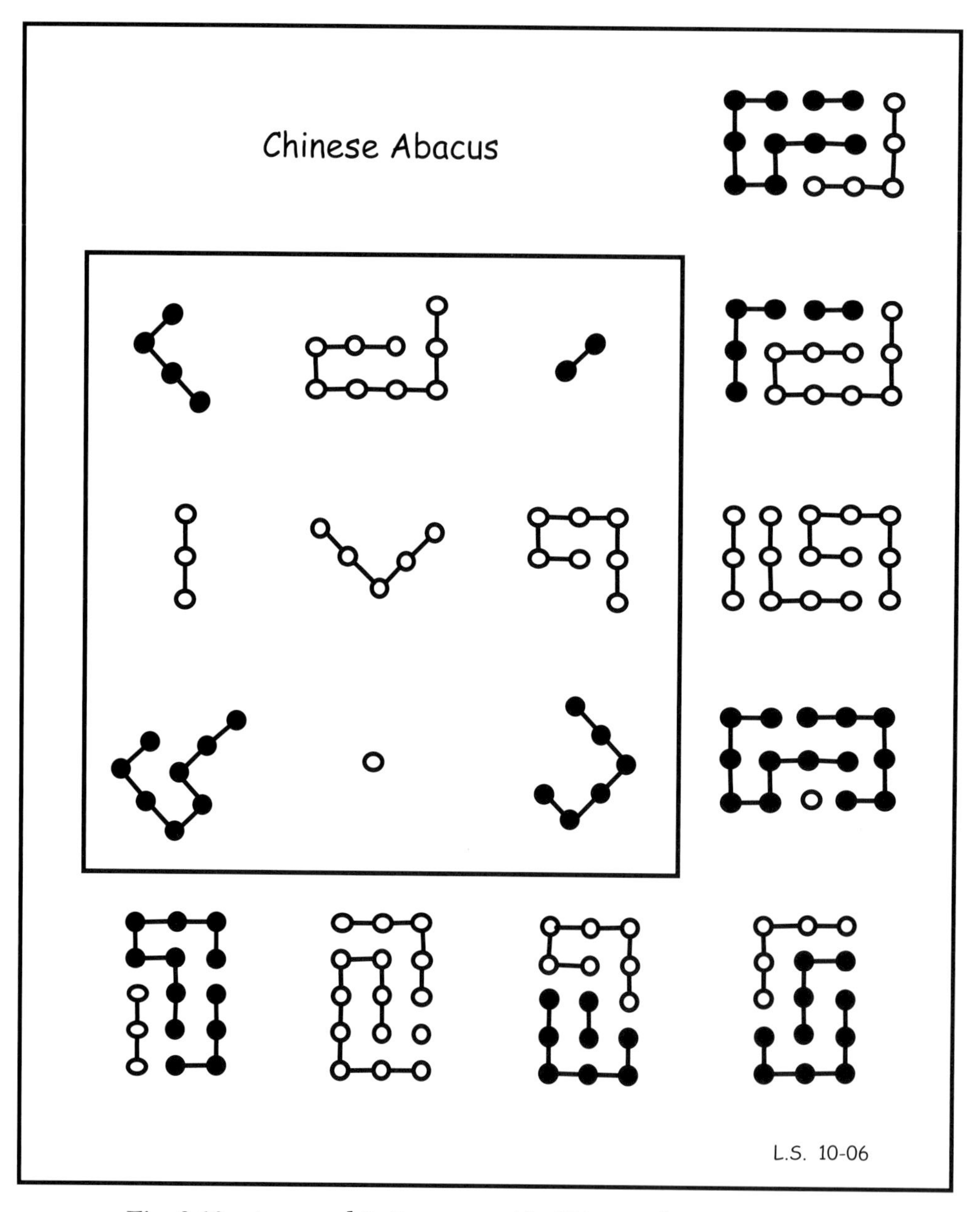

Fig. 8.12 A normal 3×3 square with Chinese abacus as target.

Part II
Geomagic Squares of 4×4

God invented the integers; everything else is the work of mantissae.

9 Geo-Latin Squares

Small is beautiful, yet the very compactness of 3×3 squares makes for stringent internal constraints that severely delimit the solutions possible. Order-4 squares are less tightly knit, for which reason they are more numerous, as well as richer in variety. Specimens were sought in the same ways as hitherto: (1) computer searches for polymagic types, and (2) hand constructions using algebraic formulae as templates. In comparing squares brought to light by the two methods, at first, the finds of the computer seem to outshine those of the pencil. In the sequel, however, the template technique yields results that exceed every expectation.

Unlike order-3, for which Lucas's square offers the only candidate, there exists a plethora of non-trivial 4×4 formulae that may be used as templates. The latter are not *general* formulae, but rather generalizations of certain subsets or special types of 4×4 numerical magic squares. The mathematical properties of these subsets need not concern us here, our interest lying solely in the use of these algebraic squares for designing 2-*D* magic squares. Many such formulae are based on *Latin squares*.

We remind ourselves that by a Latin square of order N, we refer to a square of N^2 entries composed of N distinct elements, each occurring exactly once in every row and column. Figure 9.1 shows a so-called *diagonal* Latin square of size 4×4, the elements of which are letters. Diagonal Latins are those in which each of the N distinct elements again appears exactly once along the two main diagonals. Interpreting the letters as algebraic variables, Figure 9.1 can thus itself be treated as a formula for a certain kind of very trivial 4×4 numagic square.

Since the same 4 elements occur in every straight line, we could replace each distinct letter in Figure 9.1 with *any* distinct geometrical shape, to yield a trivial but fully magic geometric square, or in other words, a *substrate* (as explained in section 4). The target in this 'geo-Latin' substrate could then be *any* spatial configuration of the 4 shapes chosen. In particular, by choosing 4 same-height rectangles of different lengths, the target can be a rectangle formed by concatenating the 4 pieces *in any order*, a property that will prove useful in the step to follow.

A	B	C	D
C	D	A	B
D	C	B	A
B	A	D	C

Fig. 9.1 A 4×4 diagonal Latin square.

As in detrivializing 3×3 substrates, a pattern of keys and keyholes is now needed that will modify these geo-Latin pieces so as to yield 16 distinct shapes. This is analogous to the task of looking for a pattern of $+x$'s and $-x$'s that could be added to the Latin square so as to yield 16 distinct entries, while preserving a constant sum in every row, column, and diagonal. However, the fact that A, B, C, and D each occur 4 times means that at least two distinct variables will be needed. A few trials with pencil and paper came up with the pattern of a's and b's in Figure 9.2, which can thus be interpreted as a formula for a certain subset of non-trivial numagic squares. The advantage of choosing pieces that can tile the target in any order now means that pieces assigned keys will be able to marry with those assigned matching keyholes, irrespective of the particular detrivializing pattern chosen.

$A+a$	$B-a$	$C+b$	$D-b$
$C-b$	$D+b$	$A-a$	$B+a$
$D-a$	$C+a$	$B-b$	$A+b$
$B+b$	$A-b$	$D+a$	$C-a$

Fig. 9.2 Every entry is unique.

Let a be a small half-circle and b a small isoceles triangle. Then replacing A, B, C, D with same-height rectangles of length 1, 2, 3, 4, respectively, we produce the geomagic square in Figure 9.3, the target of which (not shown) is a rectangle of length 1 + 2 + 3 + 4 = 10.

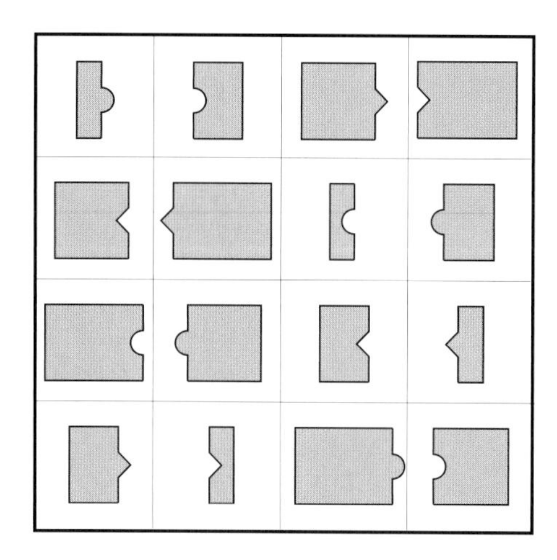

Fig. 9.3 A geometric version of Figure 9.2.

As before, the target shape could be changed to a circle or regular decagon, the pieces then becoming segments showing angles of 1, 2, 3, and 4 tenths of 360°. The result would be a 4×4 counterpart to the 3×3 square of Figure 4.9 on page 7, the keys and keyholes of which, remember, may be seen as geometric variables standing for any potential shape. Figure 9.3 is thus again a blueprint for an entire family of geomagics, in many of which the principle of construction may become difficult to detect, as it does in 'Magic Crystals' of Figure 4.10 on page 8.

Figure 9.3 has perhaps a certain austere beauty, but its artless target formations are no aesthetic match for the cunning tessellations found in many computer-discovered squares. Compare for example Figure 9.4, one of four polymagic squares using 16 hexominoes, in which the target is a 4×6 rectangle. The complete set of four can be seen in Figure 9.10.

The rectangular targets of Figure 9.3 are formed by a linear chain of 4 pieces. Figure 9.5 shows a variant scheme arrived at after some doodling. A square target is formed by 4 pieces in a *closed* chain: A abuts B abuts C abuts D abuts A. The rigid order in which A, B, C, and D now occur makes it necessary that the links in this chain are *reversible*: every piece can be flipped about an axis of bilateral symmetry (shown dotted) so as to switch hands with its two immediate neighbours.

Placing these shapes as in the Latin square on page 25 (Figure 9.1), the resulting geo-Latin substrate could be detrivialised with the key/keyhole pattern of Figure 9.2, but the outcome turns out to be unusable because of the fixed order in which the new pieces must be assembled. Those occupying the main diagonals would be unable to tile the target, as a trial will confirm. This is a warning that not every formula can be used as a template without further consideration. Figure 9.6 shows an alternative pattern that will work, again, arrived at after some trial and error. It is another non-trivial algebraic square. Some entries include both a and b.

Fig. 9.4 A 4×4 square using 16 hexominoes computer.

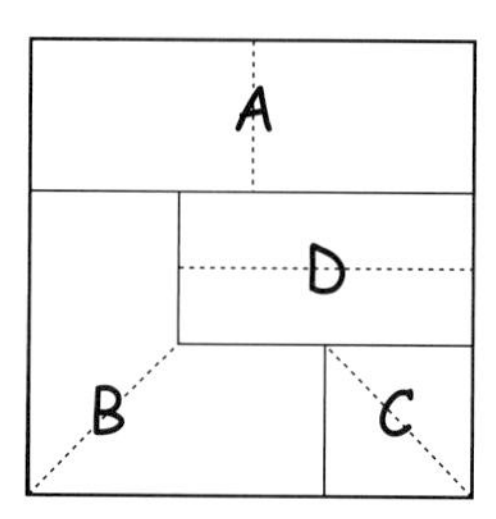

Fig. 9.5 The 4 shapes form a closed cycle.

$A+a$	$B-a+b$	$C+a-b$	$D-a$
$C+b$	D	A	$B-b$
$D-a-b$	$C+a$	$B-a$	$A+a+b$
B	$A-b$	$D+b$	C

Fig. 9.6 An alternative de-trivialised latin square.

Assigning a small square (monomino) and rectangle (domino) to a and b respectively, now results in the polymagic square of Figure 9.7. Note the identical piece

Fig. 9.7 A geometrical version of Figure 9.6.

layout in every target, a feature that can be employed to interesting effect, as we shall see later.

Again, the visual logic of Figure 9.7 may be compelling, but the unvarying target assembly is unsatisying in comparison with Figure 9.8, for example, which is another computer-discovered polymagic square.

Here, the desire for a square target dictated the choice of piece sizes. Every magic line contains three hexominoes and one heptomino. $3 \times 6 + 7 = 5 \times 5$, the area of the square target. But things can work the other way around. In an earlier chapter it was noted that there exist no 3x3 geomagics using nine pentominoes. Figure 9.9 goes some way to make up for this injustice. Here, four tetrominoes combine with the full set of twelve pentominoes to provide the 16 pieces employed. It is a matter of regret that, despite every attempt to discover a more pleasing result, the target is an incomplete rectangle.

10 4×4 Nasiks

The template method can also be used to create a 4×4 nasik square. Glancing again at the Latin square in Figure 9.1, we see that for the broken diagonals to become magic would require $2A + 2B = 2C + 2D = A + B + C + D$, or $A + B = C + D$. Rearranging the piece lengths in Figure 9.3 to $A = 1$, $B = 4$, $C = 2$, and $D = 3$, which achieves this, would,

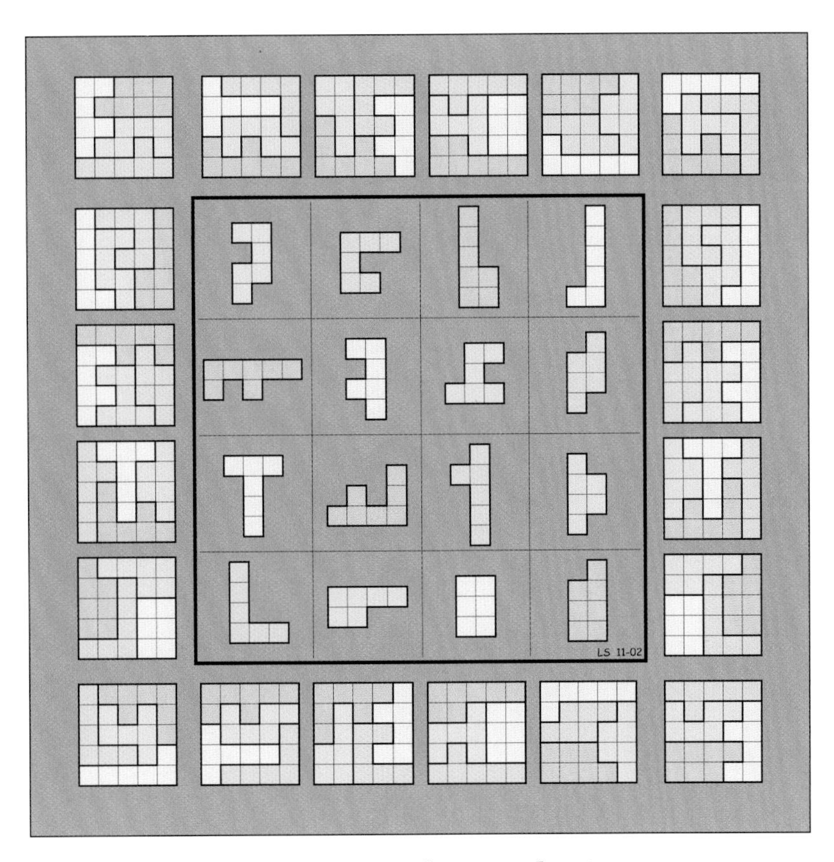

Fig. 9.8 A computer-discovered 4×4 square.

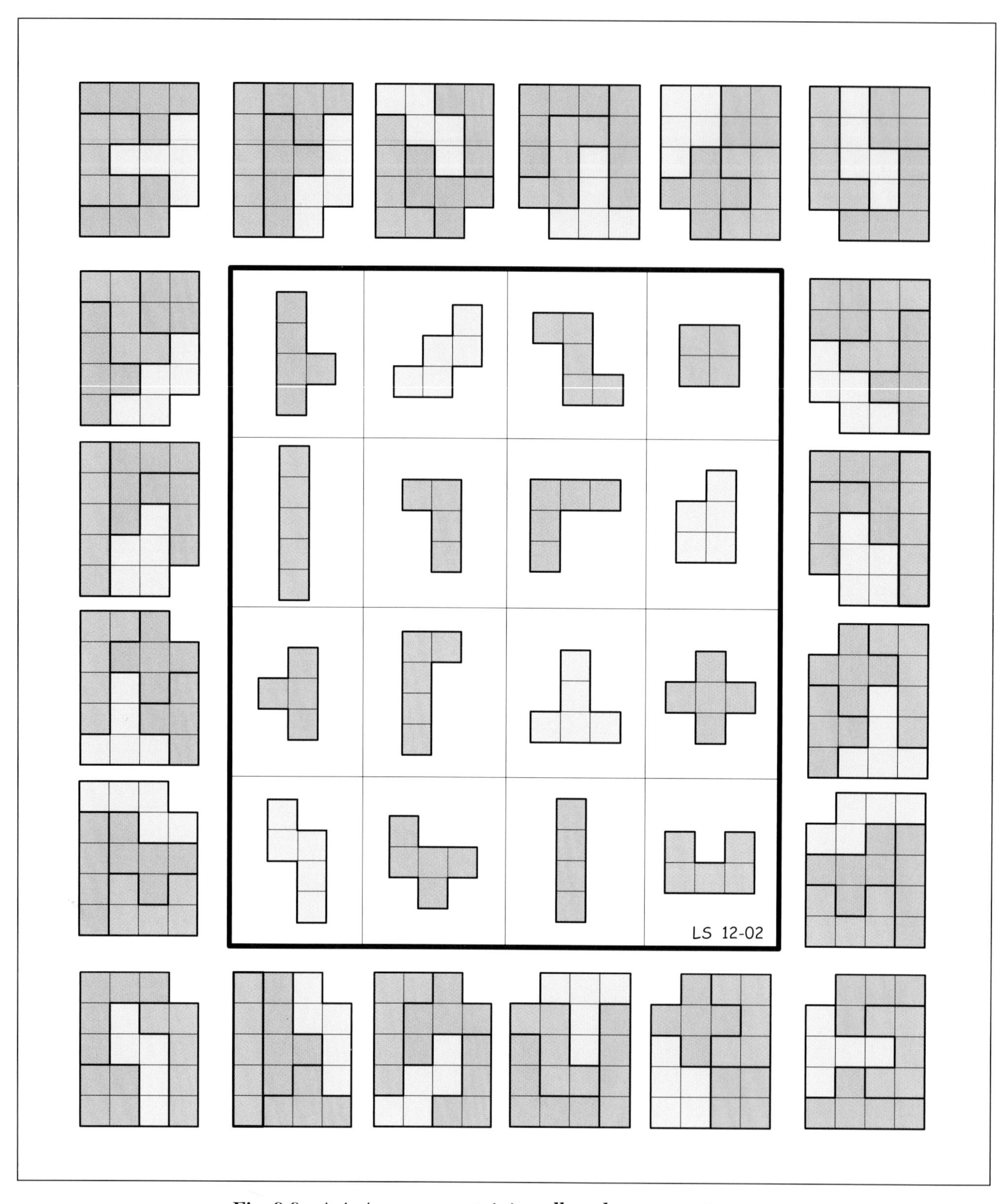

Fig. 9.9 A 4×4 square containing all twelve pentominoes.

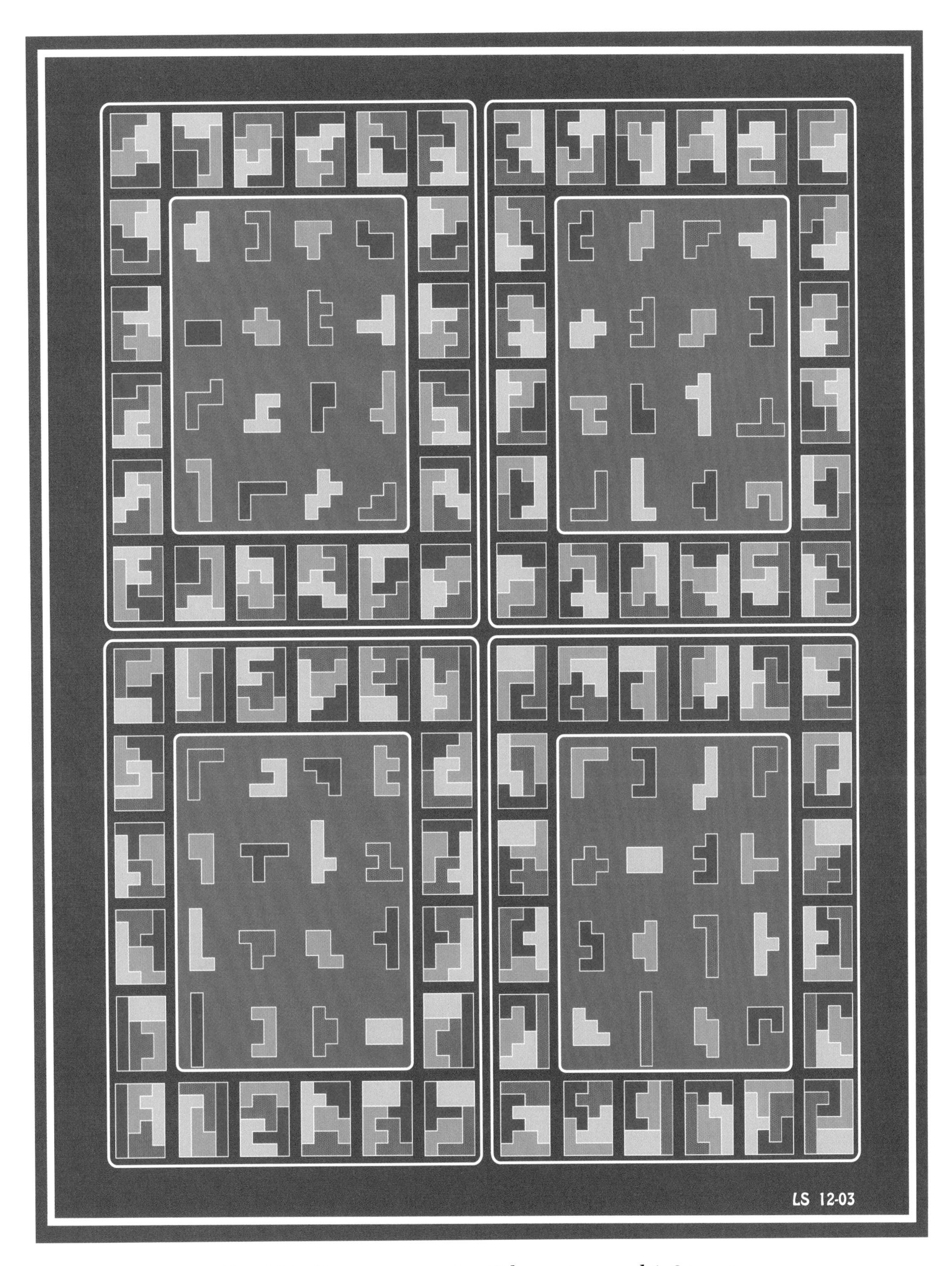

Fig. 9.10 Four squares using 16 hexominoes and 4×6 target.

therefore, result in a 4×4 nasik square, provided a new key/keyhole pattern can be found that will again detrivialize the Latin square without destroying its nasik property, as both of the previously used patterns in Figures 9.2 and 9.6 in fact do. Such nasik-preserving patterns can indeed be found, but in every case tried, the resulting set of pieces were unable to tile the target. Two keys on one piece cannot marry with their corresponding keyholes because *both* of the latter turn out to occupy another single piece.

There exist however Latin squares other than Figure 9.1. Figure 10.1 shows a *non-diagonal* 4×4 Latin square.

A	B	C	D
C	D	A	B
B	A	D	C
D	C	B	A

Fig. 10.1 A non-diagonal Latin square

Inspection shows that this becomes nasik when $A + D = B + C$, or $D = B + C - A$, the magic sum then becoming $2B + 2C$. Figure 10.2 shows a pattern of variables that detrivialises this square without interfering with its nasik property: every row, column, and diagonal sums to zero.

a	$-a$	$-b$	b
$-a$	a	b	$-b$
b	$-b$	$-a$	a
$-b$	b	a	$-a$

Fig. 10.2 Every row, column, and diagonal sims to zero.

Combining the nasik Latin with this pattern then yields Figure 10.3, a non-trivial algebraic square different again to Figures 9.2 or 9.6:

A geometrical analog of this is seen in Figure 10.4, in which *A*, *B*, and *C* are represented by same-height rectangles of length 1,3, and 2, respectively, with *a* a small half-circle and *b* a half-square. The target is a rectangle of length $2B + 2C = 10$. The piece labels A, B, . . ., P must not be confused with the variables *A*, *B*, *C* in Figure 10.3.

$A+a$	$B-a$	$C-b$	$B+C-A+b$
$C-a$	$B+C-A+a$	$A+b$	$B-b$
$B+b$	$A-b$	$B+C-A-a$	$C+a$
$B+C-A-b$	$C+b$	$B+a$	$A-a$

Fig. 10.3

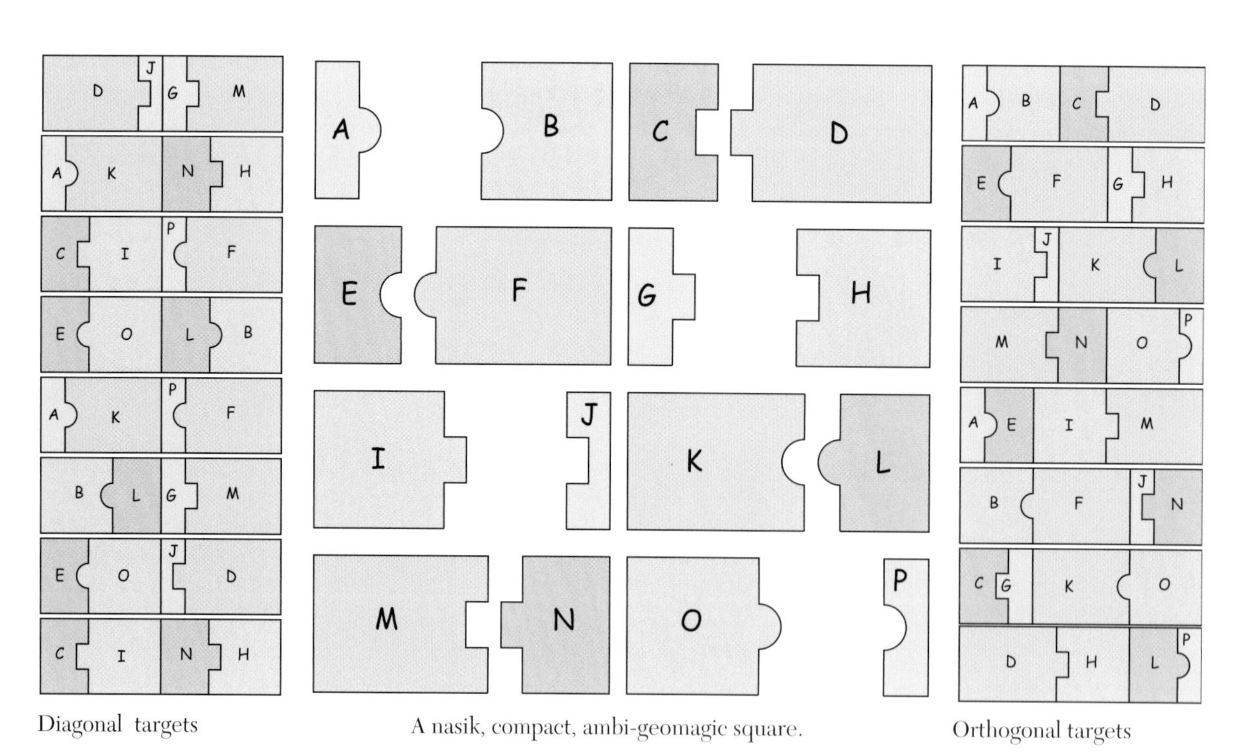

Diagonal targets

A nasik, compact, ambi-geomagic square.

Orthogonal targets

Fig. 10.4

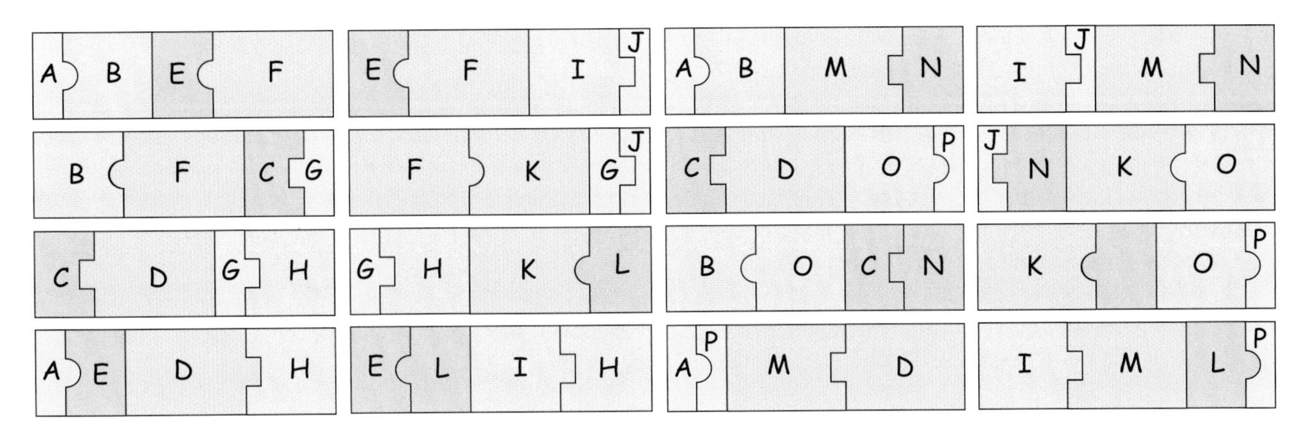

Fig 10.5 The 4 pieces occupying each of the 16 2×2 sub-squares also tile the target.

Note that the set of pieces is composed of 8 *complementary pairs: AK, BL, CI, DJ, EO, FP, GM,* and *HN*, each of which tile a rectangle of length 5, as seen in the diagonal targets, shown at left. Figure 10.4 is thus a non-trivial nasik geomagic square, but it is more besides; it is both *compact* and *ambimagic*. 'Compact' is the name I use to distinguish a 2-*D* geomagic square of order-4 in which the four entries in every 2×2 sub-square are also able to tile the target. This includes the *toroidally-connected* 2×2 squares, which is to say, those groups of 4 cells that form a 2×2 square when the top and bottom edges of the entire 4×4 array are brought together to make a cylinder that is then stretched and bent smoothly in a circle until its ends meet to form a torus. Or in other words, the 2×2 subsquares that can be found when the left and right-hand columns, as well as the upper and lower rows, are regarded as adjacent. This results in a total of sixteen 2×2 sub-squares. Figure 10.5 shows the sixteen sets of four target-tiling pieces corresponding to each sub-square in Figure 10.4. As we shall see later, every numagic square of order-4 that is nasik is also compact, and vice versa, but this is not necessarily the case for 2-*D* squares. Figure 10.4 is thus noteworthy in this regard.

Figures 10.4 and 10.5 account for 16 + 16 = 32 sets of 4 target-tiling pieces, every one of them comprising two pairs, each of which complete a shorter rectangle. However, the full total of target-tiling quads is 52, the complete list being as shown (in no particular order) in Table 1.

Below we shall see that the 16 pieces in Figure 10.4 can be rearranged to produce no fewer than 528 distinct geomagic squares, 48 of which are nasik.

'Ambimagic' which is shorthand for '*a*dditive-*m*ultiplicative *bi*-magic square,' is the name I apply to a novel kind of numerical magic square in which the orthogonals (rows and columns) each sum to the same total, while the diagonals (including the broken diagonals) each *multiply* to the same *product*. Figure 10.6 shows an example of order-4 with constant sum 60, and constant product 7560.

By extension, we can say that a geometric square is ambimagic when the pieces in every orthogonal tile one target, while the pieces in every diagonal tile another.

AEOK	AGMK	ACIK	AEIM
ABKL	AFKP	ABEF	AHKN
ABMN	ADJK	ADEH	ADMP
ABCD	CGOK	EGMO	ECIO
KLOP	BELO	EFOP	JKNO
EHNO	MNOP	BCNO	DEJO
CDOP	CGIM	GHKL	BGLM
FGJK	EFGH	FGMP	BCFG
GHMN	DGJM	CDGH	IJKL
EHIL	ILMP	BCIL	EFIJ
CFIP	IJMN	CHIN	CDIJ
BFLP	BHLN	BDJL	DHLP
BFJN	FHNP	DFJP	DHJN

Table 1 The 52 target-tiling sets.

Figure 10.4 exhibits this property. As a nasik square, the four pieces in each orthogonal tile a rectangular target of length 10, as do the four pieces in every diagonal. The diagonal sets are special however, in that each is composed of two complementary pairs, both of which tile identical rectangles of length 5; see Figure 10.4 left. The target of length 10 is formed by joining the latter end-to-end. But the two length-5 rectangles need not be so joined. They could be stacked one atop the other, say, to yield a distinct target of length 5 that is twice the height of the target for rows and columns. According to our extended definition, Figure 10.4 is thus ambimagic. Once again though, for more eye-catching target mosaics we look to computer-discovered polymagic squares. Figure 10.7 shows a nasik square using 4 tetrominoes, 4 pentominoes, 4 heptominoes, and 4 octominoes arranged size-wise as in a diagonal Latin square (but now different to Figure 9.1). The target is a 4×6 rectangle of area 4 + 5 + 7 + 8 = 24. The target is also tiled by the four pieces occupying each quadrant. For a more artistic rendering of this find see Figure 16.8.

1	35	9	15
18	12	2	28
5	7	45	3
36	6	4	14

Fig. 10.6 An ambimagic square.

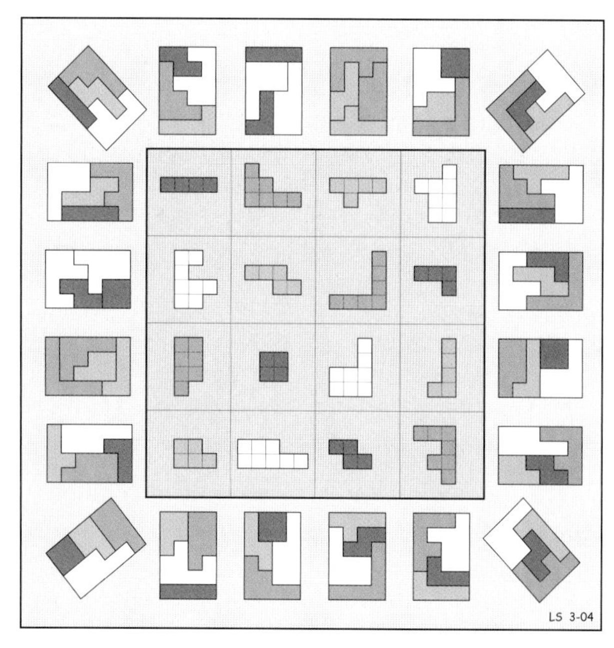

Fig. 10.7 A nasik square of 4×4.

11 Graeco-Latin Templates

A square is called *Graeco-Latin* or *Eulerian* when the two superimposed (frequently *added*) Latin squares of which it is composed result in a distinct entry in every cell. It was former practice to use Greek or Latin letters to distinguish these two components. Figure 11.1 shows an example using two *diagonal* latin squares, one shown in uppercase, the other in lowercase type. Their combination results in a square showing 16 unique entries, as required.

$A+a$	$B+b$	$C+c$	$D+d$
$C+d$	$D+c$	$A+b$	$B+a$
$D+b$	$C+a$	$B+d$	$A+c$
$B+c$	$A+d$	$D+a$	$C+b$

Fig. 11.1 A Graeco-Latin Square.

The close connection between Graeco-Latin and magic squares becomes clear from a comparison with Figure 11.2, which is a general formula[5] that describes the structure of every 4×4 numagic square. Note that the formula becomes a Graeco-Latin square when x is zero, just as variable d in Figure 11.1 can also be set to zero, in which case it need not appear.

$A+a$	$B+b$	$C+c$	D
$C+x$	$D+c$	$A+b$	$B+a-x$
$D+b-x$	$C+a$	B	$A+c+x$
$B+c$	A	$D+a$	$C+b$

Fig. 11.2 A generalization of numagic squares of order 4.

In the foregoing we have looked at geomagics based on Latin squares. It is natural to wonder whether Graeco-Latin squares might also serve as templates. At first sight it is difficult to see how. Finding a way through the difficulties presented an interesting puzzle.

Suppose we begin with a trivial geo-Latin square, G, formed by assigning distinct shapes to A, B, C, and D in Figure 11.1. Initially, it seems that four new shapes representing a, b, c and d must now be appended to these so as to detrivialize G, but without there being any corresponding keyholes to receive or "absorb" the latter.

5 Due to author. The formula can be shown to be mininal in the sence of generalizing all 4x4 magic squares more economically than any other.

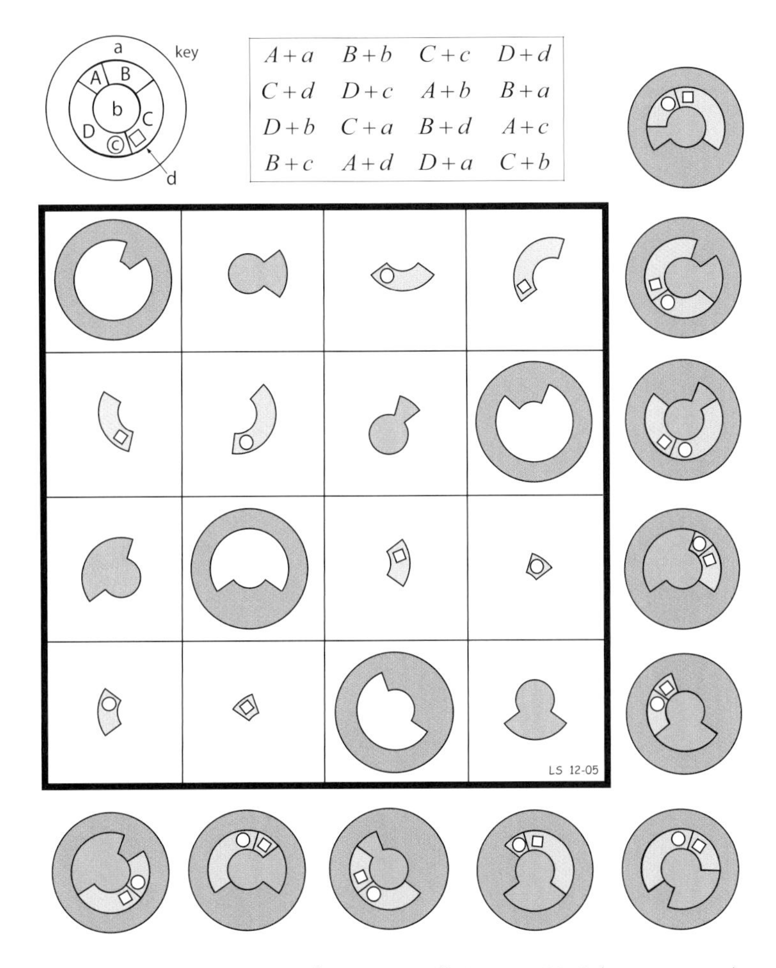

$A+a$	$B+b$	$C+c$	$D+d$
$C+d$	$D+c$	$A+b$	$B+a$
$D+b$	$C+a$	$B+d$	$A+c$
$B+c$	$A+d$	$D+a$	$C+b$

Fig. 11.3 A geometrical version of Figure 11.1 (seen at top).

The problem is thus: how then can these extra shapes be appended to the *A*, *B*, *C*, and *D* pieces so that the square remains geomagic?

We shall look at two solutions, their initial geo-Latin squares forming circular and (roughly) rectangular targets, respectively. In the first case, the final form of the target remains a circle, in the second, we are able to choose virtually any shape desired.

Figure 11.3 shows a geo-Eulerian square in which the target is a circular disc showing two adjacent holes, one square, and one circular. *A*, *B*, *C*, and *D* are represented by annular segments subtending arcs of 1, 2, 3, and 4 tenths of 360°, to result in a geo-Latin square in which the 4 pieces can be assembled *in any order* to complete an annular target. This is the ring sandwiched between *a* and *b*, the outer ring and the inner circle, as seen in the target key shown at top left in Figure 11.3.

The problem of appending magic-preserving keys-without-keyholes to these pieces is then solved by means of two separate devices:

(1) By choosing key-shapes with rotational symmetry, a property that allows them to be appended to the four different annular segments so as to preserve target circularity. These are *a* and *b*, the outer ring and inner circle.

(2) By interpreting two of the Eulerian elements as negative or excised areas. That is, the second Latin square (using small letters) is treated as shown in Figure 11.4, in which $-c$ and $-d$ correspond to the small circular and square holes, respectively.

a	b	$-c$	$-d$
$-d$	$-c$	b	a
b	a	$-d$	$-c$
$-c$	$-d$	a	b

Fig. 11.4 A latin square with negative entries.

The latter serve to detrivialize the annular segments, whose ability to be assembled in different orderings means that these two holes can always be arranged to appear in a fixed relation to each other in every target. Note that $-c$ and $-d$ are not accompanied by their positive counterparts, c and d. This concludes our glance at the circular target case.

Figure 11.5 shows a quite different geo-Eulerian square that is based on a geo-latin square using rectangular rather than annular pieces. The Eulerian template is now Figure 11.6, which is Figure 11.1 with $d = -c$, a trick that provides any key represented by c with a matching keyhole, and thus overcomes one difficulty at a stroke. As in previous cases, A, B, C, and D are now replaced by same-height rectangles of length 1 (red), 2 (green), 3 (blue), and 4 (yellow), the target formed by these being a rectangle of length 10. Again, the ability to concatenate these pieces in any desired order means that the round and square keys replacing a and b can always be manoeuvered so as to appear at opposite ends of the target, shown in skeletal form at the top in Figure 11.5. Variable c is the triangular key that marries with its matching keyhole in every target.

A more attractive elaboration of this principle is seen in Figure 11.7. *Pythagoras* is the title of a long-running, deservedly popular mathematics magazine aimed at school children in The Netherlands. 'Magie van Merlijn' or 'Merlin's Magic,' which appeared in its pages, attempted to be didactic as well as entertaining. To this end, aside from its swords, shields, flying pennants, and portcullis, the drawing included the algebraic template of Figure 11.6 seen at top, from which readers were able to follow how the square was derived.

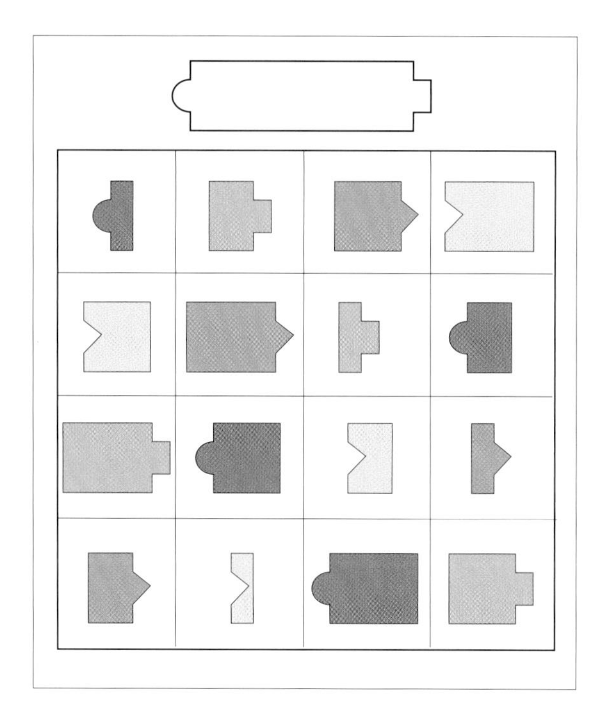

Fig. 11.5 A geo-Eulerian square.

$A+a$	$B+b$	$C+c$	$D-c$
$C-c$	$D+c$	$A+b$	$B+a$
$D+b$	$C+a$	$B-c$	$A+c$
$B+c$	$A-c$	$D+a$	$C+b$

Fig. 11.6 A Eulerian template.

Before going further, there is a subtle point involved here that deserves examination. As we have seen, Figure 11.6 is Figure 11.1 with $d = -c$. Could we not go a step further in the same direction by creating a further template in which b is set equal to $-a$, as well? The result would then seem to give rise to a distinct type of geo-Eulerian square.

The answer is yes. In fact, we already did so. It is Figure 9.2 on page 25, which was the detrivialized *Latin* square forming the template for Figure 9.3. This raises a fine distinction: whether we wish to regard Figure 9.3 as a detrivialized geo-latin square (as I do), or as a genuine geo-eulerian square (which I don't). However, on reflection we see that the addition of one (appropriate) latin square to another is in fact one way of detrivializing them both at once, so that both interpretations may be regarded as legitimate. The issue is thus perhaps academic, but I judged it worth bringing to the reader's attention.

In any case, there is a further reason for taking a closer look at Figure 11.5, which would be unremarkable were in not for the fact that the keys replacing a and b can be of almost *any* shape. Figure 11.8, or Indian Reservation, illustrates this property by combining two shapes for a and b that completely envelope the remaining rectangular part of the target and then abut each other so as to complete a wigwam, or equilateral triangle.

A is thus the small rectangular area immediately under the vertical slot in the (red) top left hand piece, the remainder of the piece (i.e., the (red) left- and right-hand lower halves of the target triangle) being the key a. In the piece to right of this is a similar (green) rectangle of length 2 above the slot, appended to which is the (green) top half of the triangle target, or key b. Key/keyhole c is a very small triangle (pointed tip) appended to the blue height-3 rectangle in the third top row cell, and excised from the yellow height-4 rectangle in the fourth top row cell (incised tail). Clearly, the triangular shape of the target is an entirely arbitrary choice that could be substituted for by any desired alternative. The interior rectangle could even be tucked away in a far corner as a small, visually insignificant component of a far more elaborate target.

I fear that this tortuous description will do little to encourage enthusiasm for Graeco-Latin based geomagics. Given the prominence of Graeco-Latins in the magic square literature however, a brief comment on the

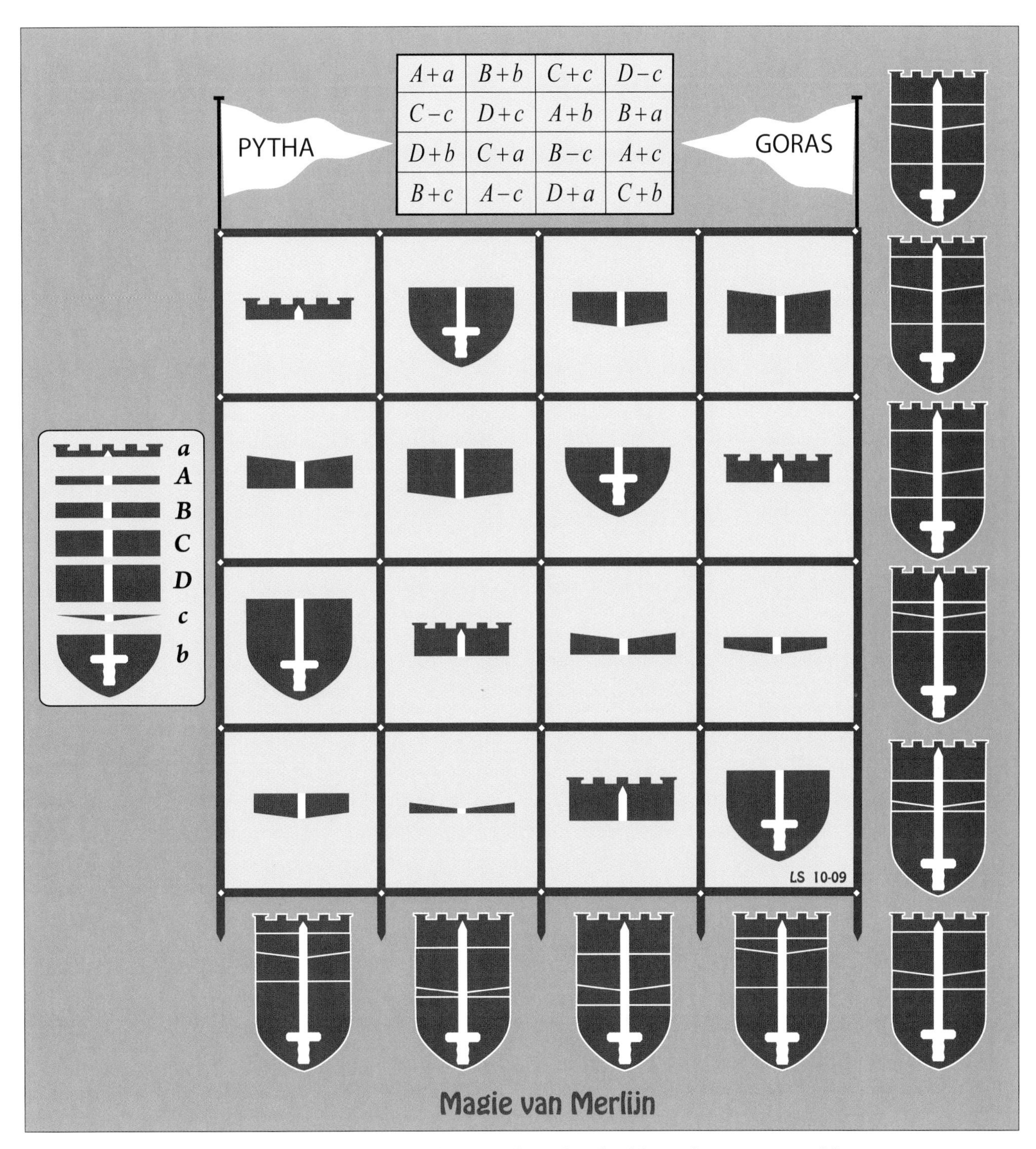

Fig. 11.7 'Merlin's Magic' The sword on the shield is, of course, 'Excalibur'.

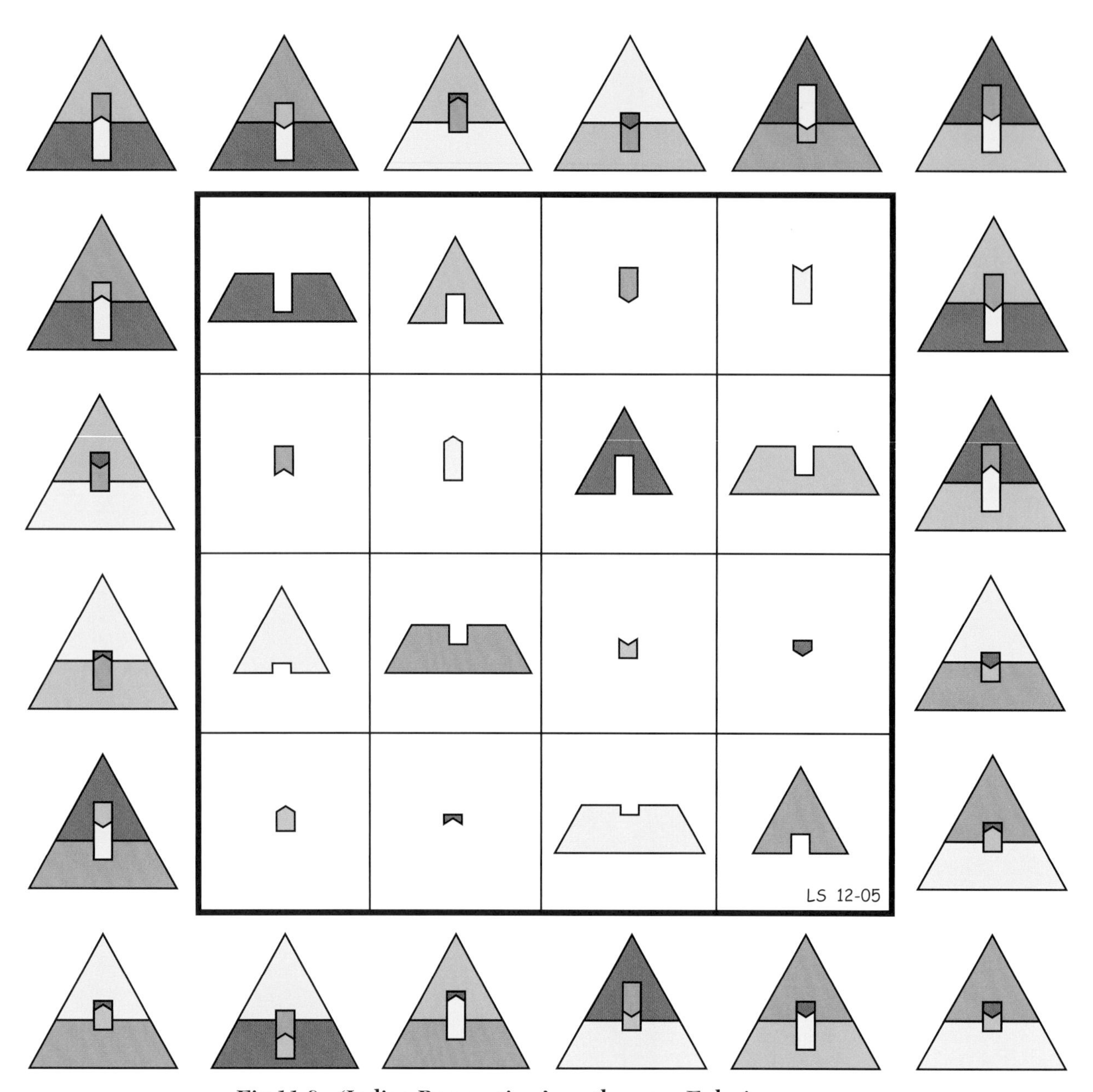

Fig 11.8 'Indian Reservation,' another geo-Eulerian square.

problems and opportunities they present seemed an appropriate topic to be raised in this account, for all its tedious character.

12 Uniform Square Substrates

We have looked at a few examples of 4×4 geomagics based on Latin squares. The process of construction was the same in each case: from an initial Latin square we produce a geo-Latin substrate to which keys can be added and/ or keyholes excised so as to yield a non-trivial geomagic square. Are there, perhaps, other substrates we might employ instead?

There is at least one avenue to explore, although at first sight it looks unpromising. Moreover, it seems almost a retrograde step. I refer to substrates that are modelled on *degenerate* Latin squares, which is to say, those in which elements repeat. Figure 12.1 shows some degenerate versions of the Latin square in Figure 9.1 (reproduced at left).

It is unnecessary to explore every case here, a pursuit that inquisitve readers may find rewarding. There is however one special instance of a degenerate Latin that will repay examination. It is the least propitious of them all: a uniform square of 16 identical elements; see Figure 12.2. A substrate corresponding to this square is therefore a uniform array of 16 identical shapes.

A	*B*	*C*	*D*
C	*D*	*A*	*B*
D	*C*	*B*	*A*
B	*A*	*D*	*C*

Latin Square

A	*B*	*C*	*C*
C	*C*	*A*	*B*
C	*C*	*B*	*A*
B	*A*	*C*	*C*

D = C

A	*B*	*C*	*B*
C	*B*	*A*	*B*
B	*C*	*B*	*A*
B	*A*	*B*	*C*

D = B

A	*A*	*C*	*C*
C	*C*	*A*	*A*
C	*C*	*A*	*A*
A	*A*	*C*	*C*

D = C, B = A

A	*B*	*B*	*B*
B	*B*	*A*	*B*
B	*B*	*B*	*A*
B	*A*	*B*	*B*

D = C = B

Fig. 12.1

A	*A*	*A*	*A*
A	*A*	*A*	*A*
A	*A*	*A*	*A*
A	*A*	*A*	*A*

D = C = B = A

Fig. 12.2 A uniform array.

A detrivializing pattern of variables is now required to produce a template from Figure 12.2. Again, relying on trial and error, with a little patience it is not difficult to come up with candidate patterns. As it happened however, this was unnecessary, several such squares being already present among my notes. This was the fruit of some earlier work on numerical magic squares. A brief digression explaining the origin of this material will assist in understanding the remarkable geomagic squares that it brought to light.

13 Dudeney's 12 Graphic Types

It is a well-known fact, first established by Bernard Frénicle de Bessy in 1693, that the number of 'normal' 4×4 magic squares that can be constructed using the natural numbers 1, 2, . . . , 16 is 880. Writing more than two hundred years later in *The Queen* for January 1910, H.E. Dudeney, the famous English puzzlist, published a system for dividing the 880 normal squares into twelve numbered "Graphic Types," depending on the twelve different patterns in which the eight so-called 'conjugate' or 'complementary' pairs, 1 and 16, 2 and 15, . . , 8 and 9, are found to occur. The same classification, including a table showing the number of normal squares belonging to each Type, appears in his well known *Amusements in Mathematics*[12]. The quaint old-fashioned diagrams in which lines link the complementary pairs have held a peculiar fascination for me from the very first moment I saw them; see Figure 13.1 and Table 2.

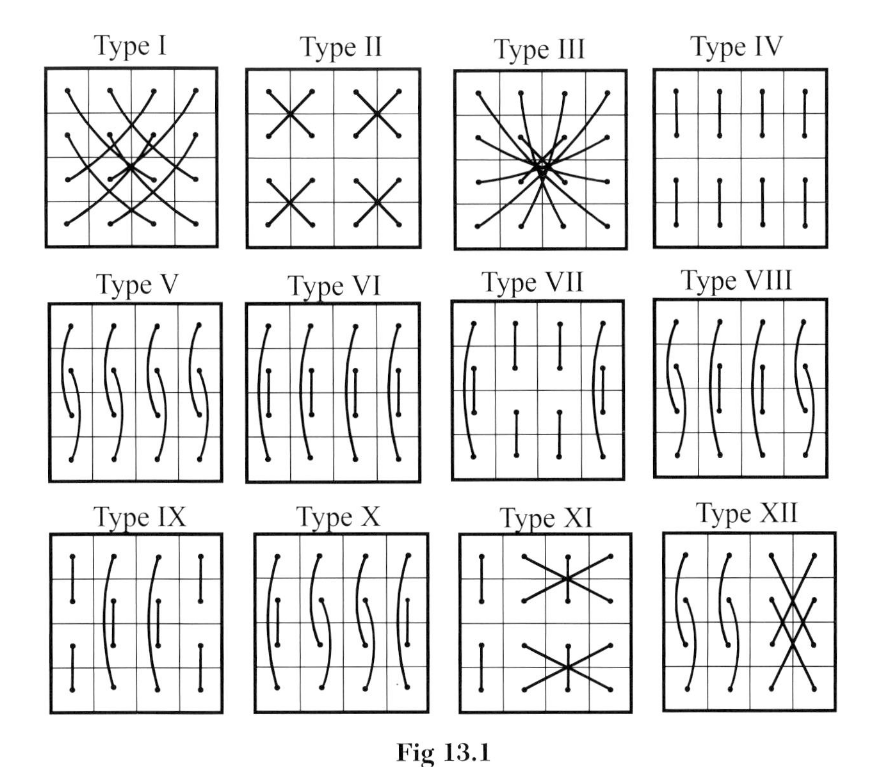

Fig 13.1

Type	Total
I	48
II	48
III	48
IV	96
V	96
VI	304
II	56
VIII	56
IX	56
X	56
XI	8
XII	8
	880

Table 2

I

pq	$\bar{p}s$	$\overline{rs}$	$\bar{q}r$
$\overline{ps}$	$p\bar{q}$	qr	$\bar{r}s$
rs	$q\bar{r}$	$\overline{pq}$	$p\bar{s}$
$\overline{qr}$	$r\bar{s}$	ps	$\bar{p}q$

II

pq	$\bar{r}s$	$\overline{ps}$	$\bar{q}r$
$r\bar{s}$	$\overline{pq}$	$q\bar{r}$	ps
$\bar{p}s$	qr	$p\bar{q}$	$\overline{rs}$
$\overline{qr}$	$p\bar{s}$	rs	$\bar{p}q$

III

pq	$\overline{qr}$	$\overline{ps}$	rs
$\bar{q}r$	$\bar{p}q$	$\bar{r}s$	$p\bar{s}$
$\bar{p}s$	$r\bar{s}$	$p\bar{q}$	$q\bar{r}$
$\overline{rs}$	ps	qr	$\overline{pq}$

IV

pq	$\bar{r}s$	$\overline{ps}$	$\bar{q}r$
$\overline{pq}$	$r\bar{s}$	ps	$q\bar{r}$
$p\bar{q}$	$\overline{rs}$	$\bar{p}s$	qr
$\bar{p}q$	rs	$p\bar{s}$	$\overline{qr}$

V

pq	$\overline{qr}$	$\bar{p}q$	$\bar{q}r$
$r\bar{s}$	$\bar{p}s$	$\overline{rs}$	ps
$\overline{pq}$	qr	$p\bar{q}$	$q\bar{r}$
$\bar{r}s$	$p\bar{s}$	rs	$\overline{ps}$

VI

pq	$\bar{p}s$	$\overline{ps}$	$p\bar{q}$
rt	$\overline{qr}$	$q\bar{r}$	$r\bar{t}$
$\overline{rt}$	qr	$\bar{q}r$	$\bar{r}t$
$\overline{pq}$	$p\bar{s}$	ps	$\bar{p}q$

VII

p	$q\bar{r}$	$\overline{pq}$	r
$\overline{ps}$	$\bar{q}r$	pq	$\bar{r}s$
ps	$\overline{qr}$	$\bar{p}q$	$r\bar{s}$
$\bar{p}$	qr	$p\bar{q}$	$\bar{r}$

VIII

pq	$\bar{r}s$	$\overline{ps}$	$\bar{q}r$
$p\bar{q}$	$\bar{p}$	$\bar{r}$	qr
$\overline{pq}$	p	r	$q\bar{r}$
$\bar{p}q$	$r\bar{s}$	ps	$\overline{qr}$

IX

pq	$\bar{r}s$	$\overline{ps}$	$\bar{q}r$
$\overline{pq}$	r	p	$q\bar{r}$
$p\bar{q}$	$\bar{r}$	$\bar{p}$	qr
$\bar{p}q$	$r\bar{s}$	ps	$\overline{qr}$

X

p	$\bar{p}q$	$\overline{qr}$	r
ps	$\overline{pq}$	$q\bar{r}$	$r\bar{s}$
$\overline{ps}$	$p\bar{q}$	qr	$\bar{r}s$
$\bar{p}$	pq	$\bar{q}r$	$\bar{r}$

XI

pq	$\overline{ppr}$	$p\bar{q}$	r
$\overline{pq}$	$\bar{r}$	$\bar{p}q$	ppr
qr	p	$\bar{q}r$	$\overline{prr}$
$\overline{qr}$	prr	$q\bar{r}$	$\bar{p}$

XII

pq	$p\bar{q}$	$\overline{ppr}$	r
qr	$\bar{q}r$	p	$\overline{prr}$
$\overline{pq}$	$\bar{p}q$	$\bar{r}$	ppr
$\overline{qr}$	$q\bar{r}$	prr	$\bar{p}$

Fig. 13.2 The formulae corresponding to each of Dudeney's 12 Types.

Being at that time engrossed with generalizations, I immediately hit on the idea of producing a set of 12 algebraic formulae that would show the internal structure of each Dudeney Type. Every Type embodies a set of relations that, taken in combination with the standard magic conditions, defines a certain subset of magic squares whose characteristics can be captured in a restricted or non-general formula. The result of this project, written up more fully in an unpublished article, "Magic Formulae," [6] dated 1980, is reproduced in Figure 13.2.

Compression is achieved by writing ab for $a+b$, and $\overline{a}$ for $-a$, the magic sum in each case being zero. Adding an appropriate constant to each cell will then yield a square having any desired magic sum. More generally, adding the same variable, say A, to every cell in, for example, the formula for Type I, results in Figure 13.3, which is thus a non-abbreviated generalization of all Type I squares, including those with a non-zero magic sum.

6 "Magic Formulae" can be found as Appendix II.

$A+p+q$	$A-p+s$	$A-r-s$	$A-q+r$
$A-p-s$	$A+p-q$	$A+q+r$	$A-r+s$
$A+r+s$	$A+q-r$	$A-p-q$	$A+p-s$
$A-q-r$	$A+r-s$	$A+p+s$	$A-p+q$

Fig. 13.3 A formula for Type I squares.

This returns us to our starting point, for Figure 13.3 is a detrivialized version of the uniform square in Figure 12.2, the magic sum being $4A$. Note that its six broken diagonals each sum to $4A$, also. This is because Dudeney's classification "Type I," although at first sight unrelated, is, in fact, indistinguishable from the classification "nasik". Specifically, it can be shown that, normal or not, every numerical 4×4 nasik square is necessarily composed of 8 complementary pairs distributed as shown in Type I, and vice versa, the sum of each pair being $2A$, or half the magic constant. In the same way, it can be shown that

$k+w+x+y+z$	$k+w-x-y-z$	$k-w+x+y-z$	$k-w-x-y+z$
$k-w+x-y-z$	$k-w-x+y+z$	$k+w+x-y+z$	$k+w-x+y-z$
$k+w-x-y+z$	$k+w+x+y-z$	$k-w-x-y-z$	$k-w+x+y+z$
$k-w-x+y-z$	$k-w+x-y+z$	$k+w-x+y+z$	$k+w+x-y-z$

Fig. 14.1 A generalization of 4×4 nasik squares.

for numerical magic squares in general, Type I = nasik = compact, a relation that, as we shall see, does not always hold for 2-*D* geomagics.

Here then was the first of twelve ready-made templates that were still there at the back of my mind some 25 years later, by which time two-dimensional squares had become the focus of interest, with numerical magic squares now occupying a back seat. A few further remarks on the background to these 12 formulae will explain my curiosity to examine a geomagic square based on Figure 13.3.

14 The 12 Formulae

It is important to realize that the expression of a generalization can take widely differing forms. Figure 13.3, for example, is simply an alternative way of writing Figure 10.3 on page 30, both of them being formulae that describe the structure of 4×4 nasik or Type I squares. Figure 14.1 shows yet a further alternative[7]. These three algebraic squares are thus mathematically synonymous, or isomorphic, being simply different expressions of the same set of intercellular relations. In short, formulae that are mathematically identical can differ enormously in outward appearance. Nevertheless, as we have seen, when it comes to using these equivalent formulae as *templates,* not only are the geomagic squares they give rise to quite different, but their magical properties can even differ from those of their parent formulae.

Likewise, any of the 12 formulae of Figure 13.2 could appear in alternate guises, some of them more elegant than others. By 'elegant', I mean more visually symmetrical, as well as more condensed or economical in symbol appearances. It was with this in mind that these 12 squares were produced. That is to say, far from being simply calculated or deduced, they are the product of a gradual process of refinement based upon trial and error, the motive behind which was primarily *aesthetic*, with the aim of culminating in an ideal, or canonical exemplar, for each Type. Candidate squares arrived at via this process then had to be tested for validity by comparing them for isomorphism with a standard formula, known to be correct. As a magic-square buff hooked on symmetry, it pleased me to strive patiently for the most elegant expression of each formula, even if the end result was of no mathematical significance.

But as chance had it, this artistic impulse led to an incidental discovery of some interest. For I came to notice that in certain cases the formulae for distinct Types could be written using the *same set* of 16 algebraic terms. The effect of this insight prompted a more systematic analysis, the result of which was decisive for the outcome. In particular, as inspection of Figure 13.2 will verify, the final version of the 12 formulae fall into four classes sharing identical entries:

Class 1 : Types I, II, III, IV, V, (and that subset of Type VI when t = s or $-s$)
Class 2 : Type VI (excluding the above subset members)
Class 3 : Types VII, VIII, IX, X
Class 4 : Types XI, XII

Properly understood therefore, Figure 13.2 embodies a visual demonstration of something that is otherwise far from obvious, namely, that the entries of a magic square belonging to a given Class may always be transposed so as to form a new square of different Type in the same Class. Given any square of Type I, for example, its entries can be rearranged in the way shown by the formulae to produce still others of Types II, III, IV, V, and VI. How many can be formed in all?

As shown by Dudeney's table in Figure 13.1, the numbers 1, 2, . . . , 16 can be placed so as to produce 48 Type I squares. Dudeney arrived at this result by enumeration, which is to say, by counting squares. It wasn't until 1938 that a paper by Rosser and Walker [13] gave a proof of the same result based upon group-theoretic arguments. This was an important advance, although the authors might have achieved more.

In fact, it is not difficult to extend Rosser and Walker's

7 This formula is of special interest in bringing to light an archetypal 4×4 magic square. Let $w = x = y = z = 1$, with $k = 0$. Now interpret entries as 4-vectors; e.g., the bottom right hand entry becomes the vector [1,1,–1,–1]. The result is a canonical 4 × 4 nasik square whose entries are the 16 vertices of a 2 × 2 × 2 × 2 tesseract centered on the origin of 4-space!

result to the 16 algebraic terms in Figure 13.3 (or its isomorphic twin, Figure 10.3). That is, it can be shown that the entries in Figure 13.3 (or 10.3) can be placed so as to yield 48 distinct Type I squares. But as just explained, the entries of any Type I square can be rearranged to produce still others of Types II - VI, the total for each again being the same as that given in Dudeney's table, with one exception: the total for Type VI is not 304 but 192, a discrepancy that is accounted for by the fact that Type VI squares can also be constructed using Class 2 entries as well as those of Class 1. Thus, adding these totals together shows that the entries in Figure 13.3 (or 10.3) can be rearranged to produce 48 + 48 + 48 + 96 + 96 + 192 = 528 distinct magic squares, rotations and reflections not counted.

How, I wondered, would a geomagic square derived from Figure 13.3 differ from that based on Fig 10.3? Would the resulting set of pieces exhibit similar properties so as to yield 528 geomagic squares? How would the enhanced visual symmetry of the new template reflect itself in the result?

15 A Type I Geomagic Square

Digging out an old copy of *Magic Formulae*, I pondered the template of Figure 13.3. The variables p, q, r, s could be allotted four distinct shapes that would become keys and keyholes on the substrate pieces represented by A. But what shape to assign the latter? And what shape should the target take? A little thought showed that the presence of a double key/keyhole on each piece means that those in the target must form a *closed circuit*, for an open chain would leave a key and a keyhole unmarried. This implied a circular or regular polygonal target. In the latter case, if $A = 1$ the magic sum is 4, which suggested a regular 4-gon, otherwise known as a *square*. Now here was a promising start.

On analogy with the hexagonal target met with earlier, this square target could now be divided radially into 4 identical segments subtending angles of $360° \div 4 = 90°$. Three possibilities present themselves, two of them yielding symmetrical pieces, shown at right in Figure 15.1.

Choosing the rightmost, the substrate *pieces* then become square-shaped, too. Assigning now 4 distinct shapes to p, q, r, and s, Figure 15.2 shows the geomagic square to emerge from these meditations. The four colors

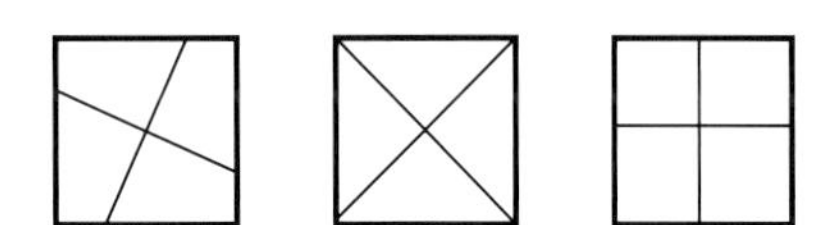

Fig. 15.1 Dividing a square into four similar pieces.

used, arranged in a Latin configuration, make it easier to follow the construction.

Here then was a nice looking geomagic square, but a glance shows that at least one important property of its template is lacking. The four pieces occupying each of the long broken diagonals, such as $p + q$, $-r + s$, $-p - q$, and $r - s$, will *not* tile the target. The two key/keyholes on any one piece must marry with the two keyhole/keys on another (its complement), which is obviously impossible. Hence, even though it is of Type I, this geometric magic square is *not* nasik.

In the realm of numerical magic squares a nasik square is always of Type I, and a Type I square is always nasik. Why should geomagics behave differently? It is because the combinative properties of numbers are not shared by geometrical forms. In the template technique we replace algebraic variables with geometrical shapes, a stratagem that owes more to abracadabra than it does to mathematics. Small wonder then if this dubious method occasionally leads to odd results. For, as with my original square based on Lucas's formula (Figure 2.4), there can be no guarantee that the pieces it yields are able to successfully tile the target, even though their combined area will be the one required.

Notwithstanding this nasik deficiency, Figure 15.2 retains most of the attributes of its template. In particular, it embraces many other quads of target-tiling pieces besides those seen in the 10 targets depicted. The square is compact, for example, a property that entails a further 16 target-tiling quads. It is natural to wonder how many there are in total.

The magic constant in the template is $4A$. The sum of the 4 algebraic terms corresponding to any quad of target-tiling pieces must therefore be the same. A computer program that examined in turn every set of 4 distinct terms occurring in the formula finds 52 that sum to $4A$. In the nasik square of Figure 10.4, every one of these corresponds to a target-tiling piece set. But not all of these will work here. There are 8 'forbidden' sets among the 52 whose corresponding pieces will not tile a square. They are those that consist of two complementary pairs having no variable in common, as shown in Figure 15.3.

It is easy to see why these sets will not tile the target in spite of possessing the appropriate area of $4A$. In the geomagic square these 8 sets are to be found occupying (1) the 4 long broken diagonals and (2) the 4 corner cells of the four embedded 3×3 squares. In the list of 52 target-tiling quads associated with Figure 10.4 (Table 1, page 31) they are the sets *AHKN*, *DEJO*, *CFIP*, *BGLM*, *ACIK*, *BDJL*, *EGMO*, and *FHNP*. The full number of target-tiling piece sets in Figure 15.2 is thus $52 - 8 = 44$.

As discussed above, a more striking attribute of the template is that its 16 algebraic terms (the Class-1 entries) can be rearranged so as to form new squares of Types

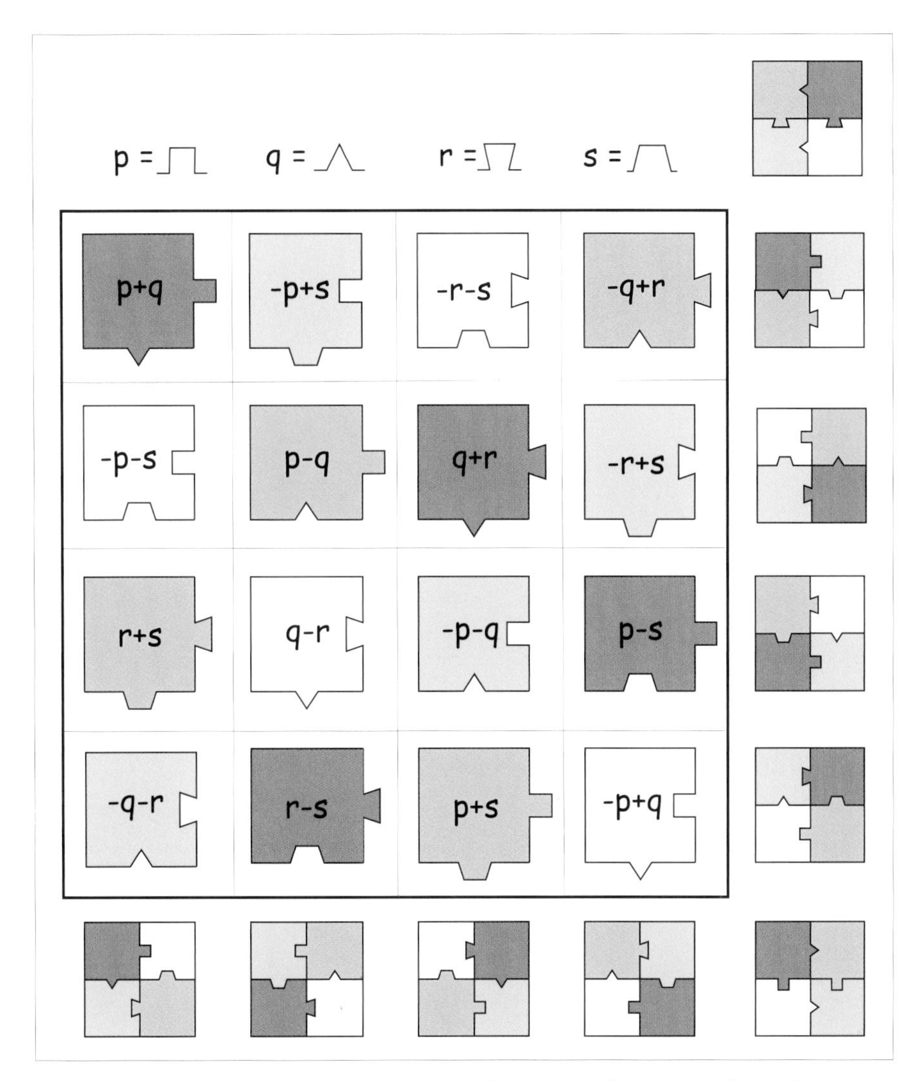

Fig. 15.2 A geomagic square based on the Type I formula.

Type	total
I	16
II	16
III	16
IV	32
V	32
VI	32
	144

Table 3

II, III, IV, V, and VI. Identical rearrangements of the 16 pieces in Figure 15.2 will thus result in new geomagic squares, *provided no forbidden set occupies a row, column, or main diagonal*, which would render the square non-magic. A computer program that generated all 528 squares in turn and then filtered out the latter cases found 144 squares remaining. The 16 pieces in Figure 15.2 will thus yield 144 distinct geomagic squares distributed over Types I – VI, the totals for each Type being as shown in Table 3.

These figures exclude rotations and reflections. Examples of geomagics corresponding to each Type are seen in Figure 15.4. Note that the Type I square shown is distinct from that appearing in Figure 15.2.

16 Self-Interlocking Geomagics.

We have seen that the set of 16 pieces derived from the template can be used to create 144 distinct geomagic squares. It was in exploring this interesting collection that a remarkable property of the Type V and VI squares came to light. As the examples in Figure 15.4 will show, the four pieces in each quadrant (e.g., $p + q$, $-p + s$, $p - q$, $-p - s$)

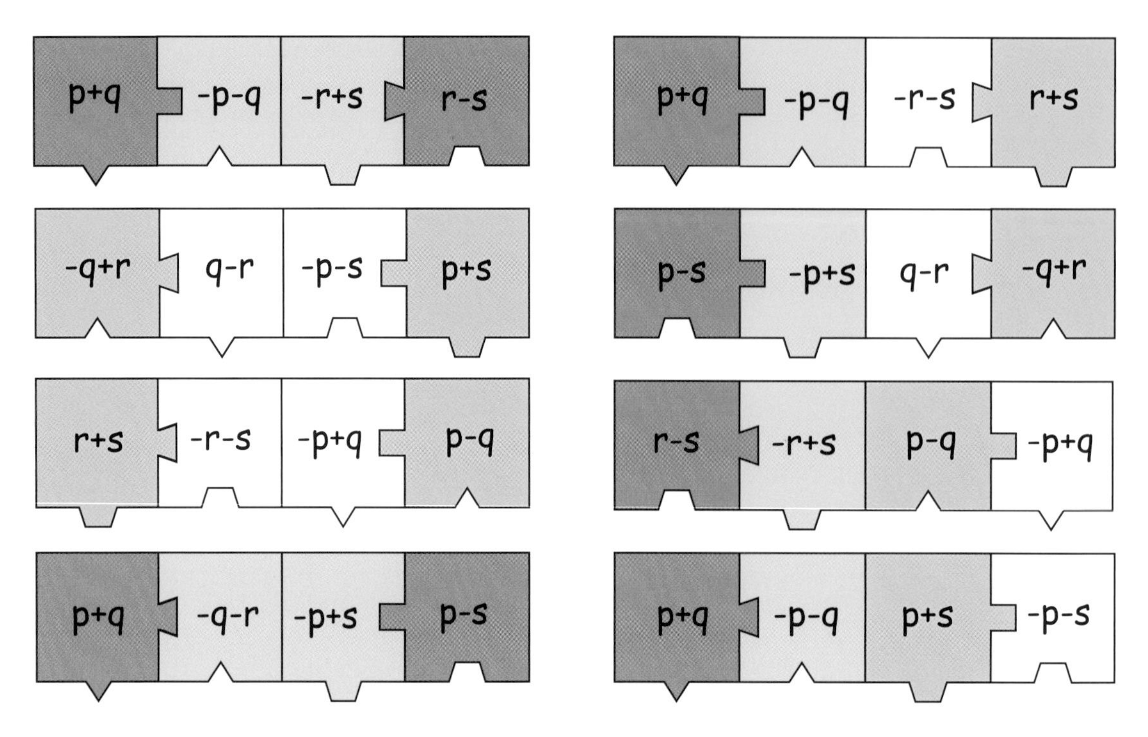

Fig. 15.3 Every quad consists of two complementary pairs having no variable in common.

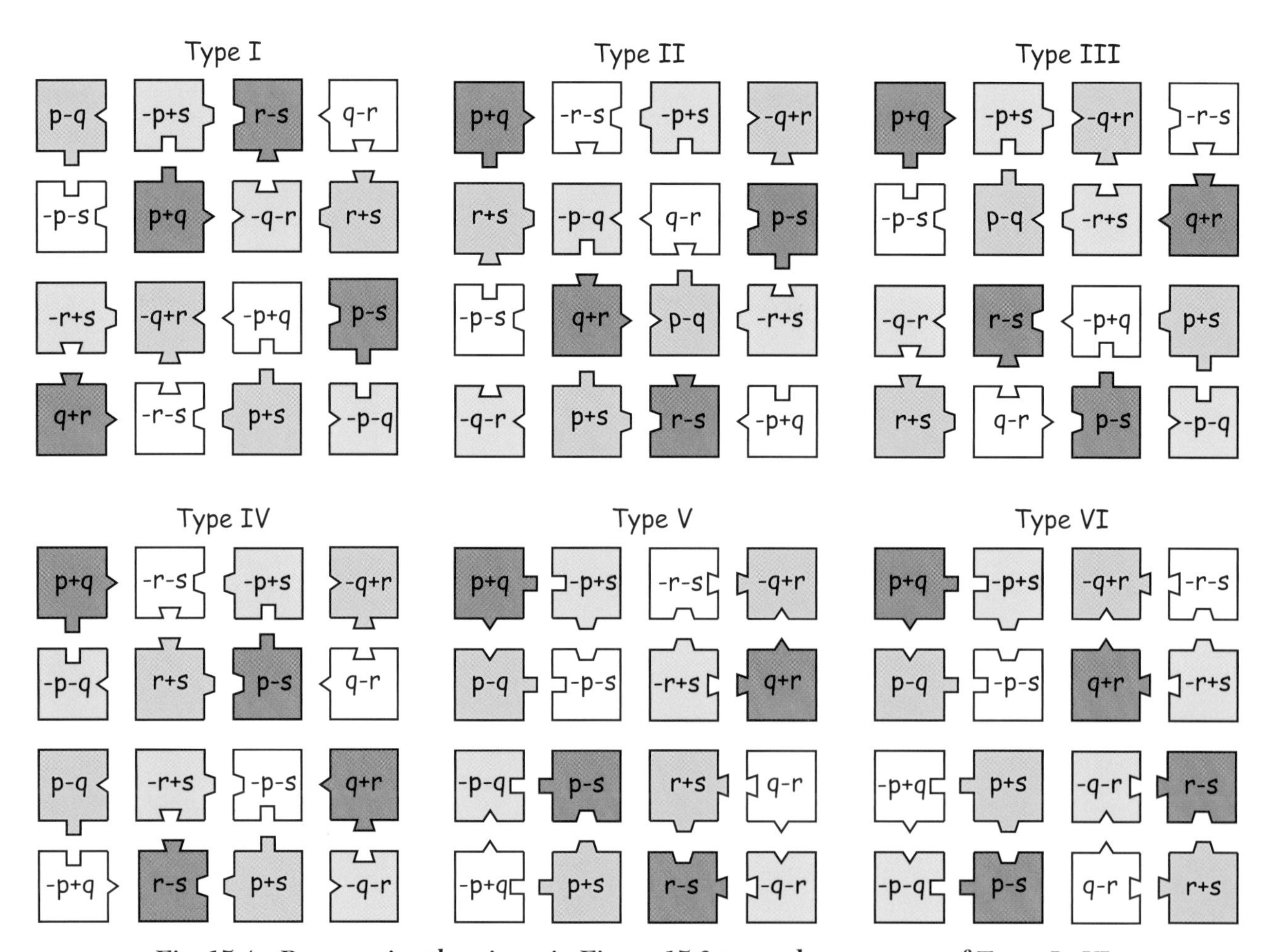

Fig. 15.4 Rearranging the pieces in Figure 15.2 to produce squares of Types I - VI.

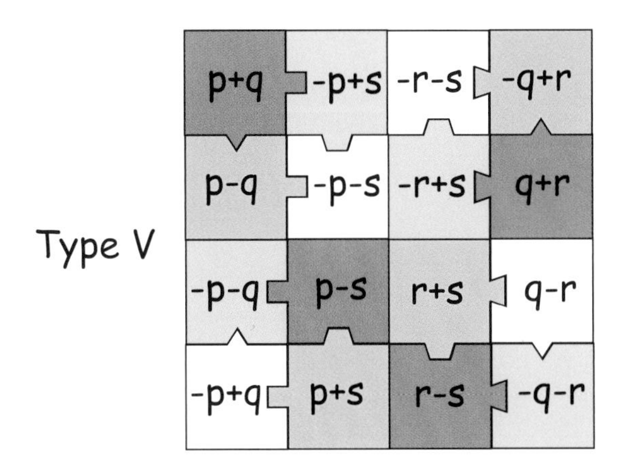

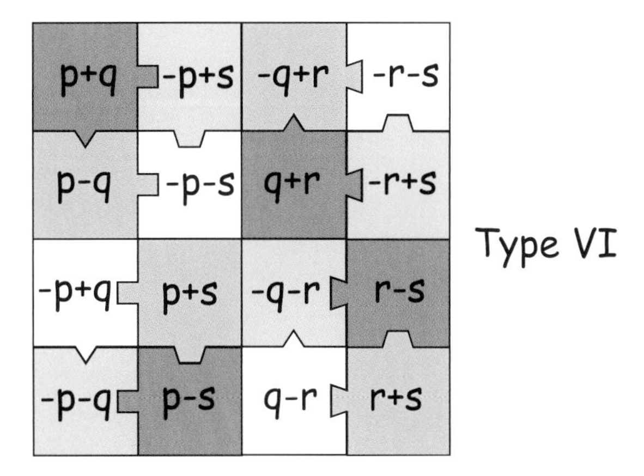

Fig. 16.1 Two self-interlocking geomagic squares.

can be squeezed together so as to form the target. These four target squares can in turn be squeezed together so as to make a larger square combining every piece. The entire set of sixteen pieces can thus be imploded without any change in their orientation so as to form a single larger square. But this means that this larger square is itself a geomagic square of Type V or Type VI, the constituent pieces of which happen to abut each other so as to form a regular quadrilateral, as seen in Figure 16.1.

Thus far we have always thought of a geomagic square as an array of separate pieces, each occupying a distinct cell. Here we have an array of shapes whose edges virtually coincide with, and thus define, their own cell boundaries. That there exist many different sets of pieces that can be placed in a 4×4 array so as to yield a geomagic square is perhaps a surprising thing in itself. It would be counted a bonus if such a set of pieces could be rearranged in some way so as to tile a square. In Figure 16.1, we are looking at a structure that achieves both of these things *at once*. For this reason, I call these geomagic squares "self-interlocking," and there are 64 of them in all, 32 of Type V and 32 of Type VI, rotations and reflections not counted.

To put the above in perspective, recall that the algebraic template of Figure 13.3 is merely an alternative expression of Figure 10.3, itself used as a template to construct the nasik square of Figure 10.4. So here we have two *isomorphic* templates, the one leading to a nasik, and the other to a non-nasik geomagic square. This again illustrates the peculiar property of the template technique already encountered, namely, that the geomagic square it yields is determined as much by the algebraic structure described by the formula, as it is by the particular set of expressions in which the formula happens to be couched. It is perhaps thus no coincidence that the "most elegant" formula has given rise to these aesthetically pleasing self-interlocking squares.

The keys and keyholes used to modify the pieces in Figure 16.1 are entirely arbitrary in shape, which means that alternatives can be used. "Magic Jigsaw Puzzle" in Figure 16.2 shows a re-expression of the Type V square in Figure 16.1 using shapes more appropriate to its title. Twenty of the 44 possible targets surround the square to complete a pleasing symmetrical ensemble. Note that the central square really does make a jigsaw *puzzle,* in that viewers can be invited to reassemble the 16 pieces so as to form a new square. There are 64 solutions, every one of them forming a distinct self-interlocking geomagic square. They are listed in Table 4. One simple solution is to switch diametrically opposed quadrants.

Curiously, the key and keyhole shapes used in 'Magic Jigsaw Puzzle' are more obviously 'arbitrary' than their originals, making both squares equally suitable as a blueprint for designing alternative versions of the same structure. Figure 16.3 shows another interpretation of the Type V square in which the keys and keyholes are monominoes and dominoes, to result in pieces, here slightly separated from each other, that are polyominoes. In this case, all 44 targets are shown. Below we shall take a closer look at exactly how the shapes of the polyominoes are arrived at.

Note that symmetries in the key/keyhole interlocks mean that the four column targets can each be tiled in three different ways. In the left-hand column target, for example, the top half of the target square formed by pieces *A* and *E* could be flipped about its vertical midline and still marry with *M* and *I*, as required.

There is an alternative way to represent polyominoes, which is by tracing paths that link the centres of their constituent square cells, as in the example using piece *A*, shown in Figure 16.4.

Such paths can often be drawn in different ways. Depending on the polyomino, forks in the path may be demanded, as in the example shown. Applying a similar approach to every piece in Figure 16.3 results, for example, in Figure 16.5, which is reminiscent of so-called 'space-filling curves' as well as designs to be found in Arabic art.

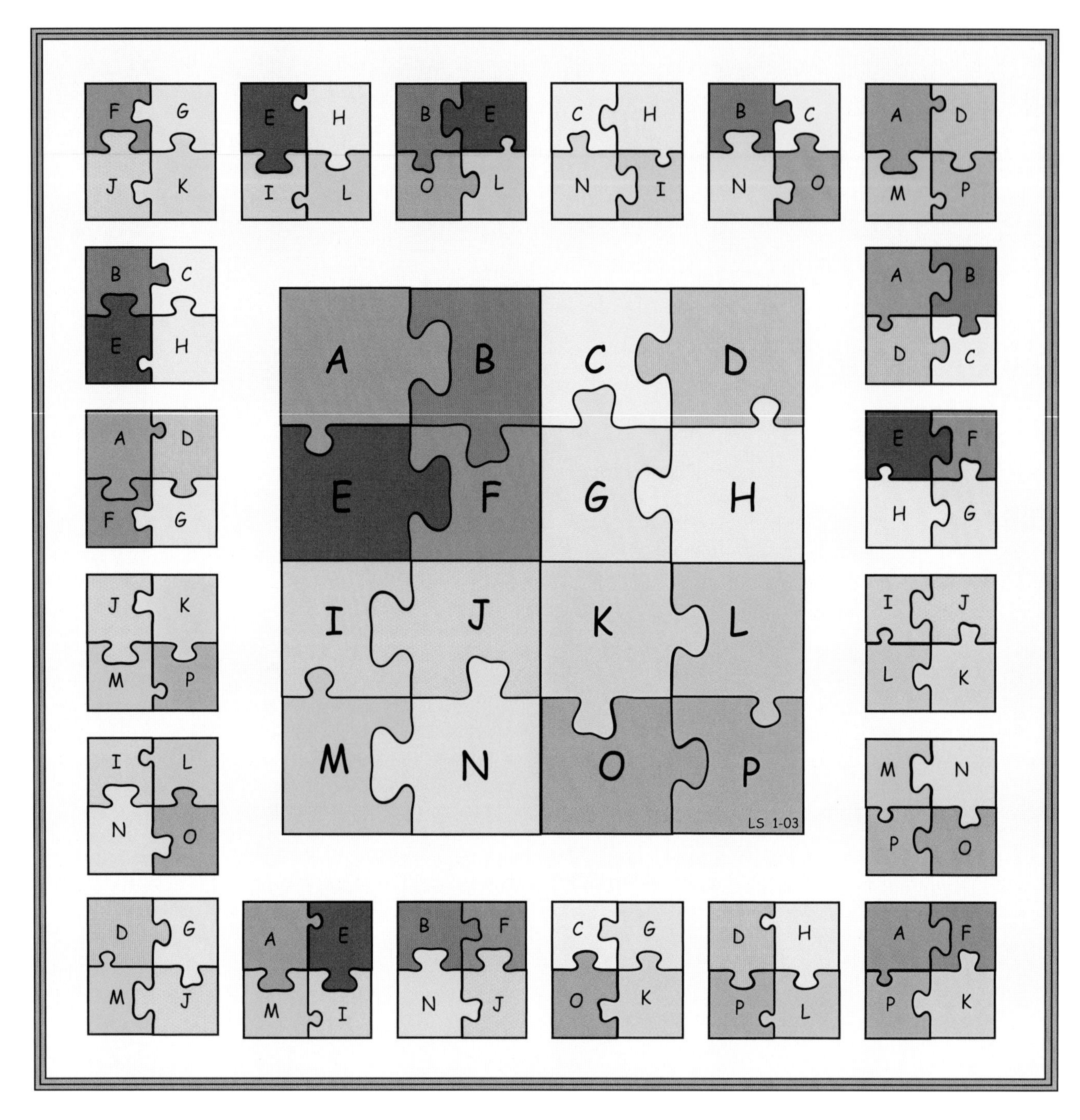

Fig. 16.2 Magic Jigsaw Puzzle.

1	2	3	4	5	6	7	8
A B C D	A B D C	A B P O	A B O P	A D B C	A D C B	A D F G	A D G F
E F G H	E F H G	E F L K	E F K L	M P N O	M P O N	M P J K	M P K J
I J K L	M N P O	M N D C	I J G H	E H F G	I L K J	E H B C	I L O N
M N O P	I J L K	I J H G	M N C D	I L J K	E H G F	I L N O	E H C B

9	10	11	12	13	14	15	16
A F D G	A F G D	A F K P	A F P K	A P B O	A P F K	A P K F	A P O B
E B H C	E B C H	E B O L	E B L O	M D N C	M D J G	M D G J	M D C N
M J P K	I N O L	I N C H	M J D G	E L F K	E L B O	I H C N	I H G J
I N L O	M J K P	M J G D	I N H C	I H J G	I H N C	E L O B	E L K F

17	18	19	20	21	22	23	24
B A C D	B A D C	B A P O	B A O P	B C A D	B C E H	B F J N	B F J N
F E G H	F E H G	F E L K	F E K L	N O M P	N O I L	E A M I	E A M I
N M O P	J I L K	J I H G	N M C D	F G E H	F G A D	H D P L	L P D H
J I K L	N M P O	N M D C	J I G H	J K I L	J K M P	C G K O	O K G C

25	26	27	28	29	30	31	32
B F N J	B F N J	B O A P	B O E L	C B A D	C B E H	C B H E	C D N M
E A I M	E A I M	N C M D	N C I H	O N M P	O N I L	O N L I	G H J I
C G O K	O K C G	F K E L	F K A P	K J I L	K J M P	G F D A	O P B A
H D L P	L P H D	J G I H	J G M D	G F E H	G F A D	K J P M	K L F E

33	34	35	36	37	38	39	40
C H B E	C O G K	C O K G	C O K G	D A G F	D C M N	D H P L	D P H L
G D F A	N B J F	N B F J	N B F J	P M K J	H G I J	G C K O	M A I E
O L N I	D P H L	I E A M	M A E I	H E C B	P O A B	A E M I	C O G K
K P J M	M A I E	H L P D	D P L H	L I O N	L K E F	F B J N	N B J F

41	42	43	44	45	46	47	48
F A D G	F A K P	F A P K	F B J N	F B N J	F B N J	F G E H	F K A P
B E H C	B E O L	B E L O	E A I M	E A M I	E A M I	J K I L	J G M D
N I L O	J M G D	N I H C	G C K O	H D P L	L P D H	B C A D	B O E L
J M P K	N I C H	J M D G	H D L P	G C O K	K O C G	N O M P	N C I H

49	50	51	52	53	54	55	56
E A I M	E A I M	E A M I	E A M I	E A M I	E A M I	E H F G	E L B O
B F N J	F B J N	B F J N	B F J N	F B N J	F B N J	I L J K	I H N C
L P H D	H D L P	C G K O	O K G C	G C O K	K O C G	A D B C	A P F K
O K C G	G C K O	H D P L	L P D H	H D P L	L P D H	M P N O	M D J G

57	58	59	60	61	62	63	64
K F A P	K F E L	K G C O	K G C O	I E A M	I E A M	I E A M	I E A M
G J M D	G J I H	J F B N	J F B N	H L P D	H L P D	L H D P	L H D P
C N I H	C N M D	I E A M	M A E I	C O K G	G K O C	K G C O	O C G K
O B E L	O B A P	L H D P	P D H L	N B F J	J F B N	J F B N	N B F J

Table 4. The 64 magic rearrangements of 'Magic Jigsaw Puzzle.'

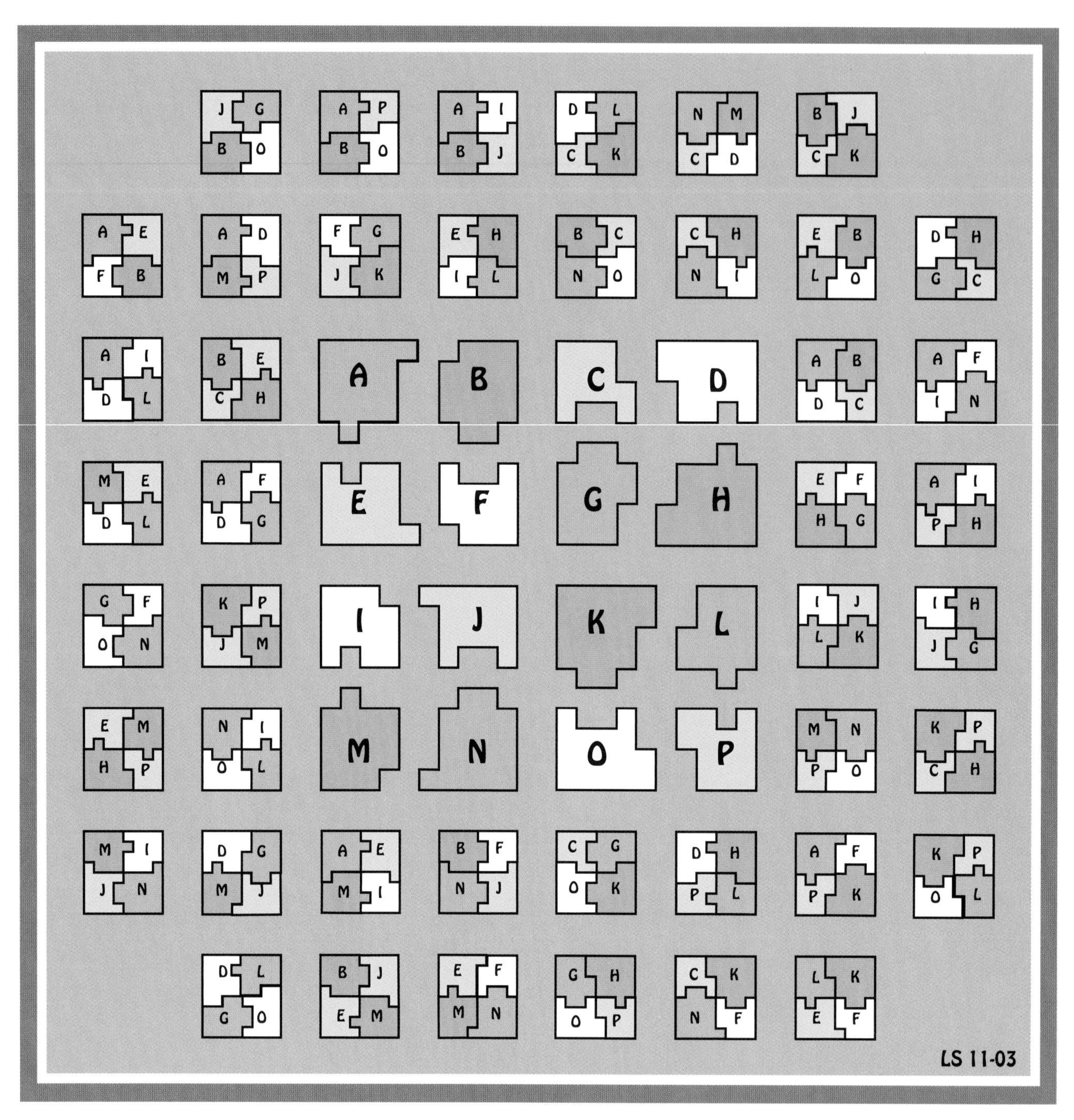

Fig. 16.3 An alternative interpretation of the Type-V formula.

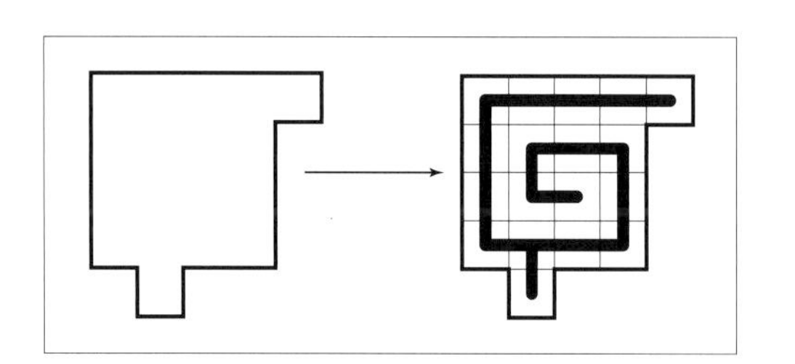

Fig. 16.4 Representing polyominoes by tracing paths.

Figure 16.6 shows the same square but now using 4 different colors distributed as in a diagonal Latin square. The central square is surrounded by 20 of its 44 targets. Here the title, 'Spatial Fugue,' was suggested by analogies with the fugal music of J.S. Bach. Harmonic alignments, both 'horizontal' and 'vertical' are the chief characteristic of counterpoint, the resultant pattern of sounds forming a temporal sequence that, in the case of the composer Bach, can truly be described as 'magic,' both in its figurative, as well as its literal sense.

Fig. 16.5 An alternative representation of Figure 16.3.

Spatial Fugue would make a nice design for a carpet, perhaps. Speaking of which, Figure 16.7 shows 'Magic Carpet V' (the fifth in a series of experiments), a square that is both pandiagonal and compact, in which the polyomino pieces are again represented by paths. The 36 targets surrounding the central geomagic square (8 for the rows and columns, 8 for the diagonals, 16 for the 2×2 subsquares, and 4 for the corners of the 4 embedded 3×3 sub-squares) are 4×6 rectangles minus two adjacent cells on one corner. Their symmetrical layout is intended to suggest a carpet design.

'Magical Alphabet,' another nasik square seen in Figure 16.8, is in fact a re-expression of Figure 10.7 in the medium of paths rather than polyominoes. The result struck me as suggestive of an exotic script.

Fig. 16.6 'Spatial Fugue,' a study in counterpoint.

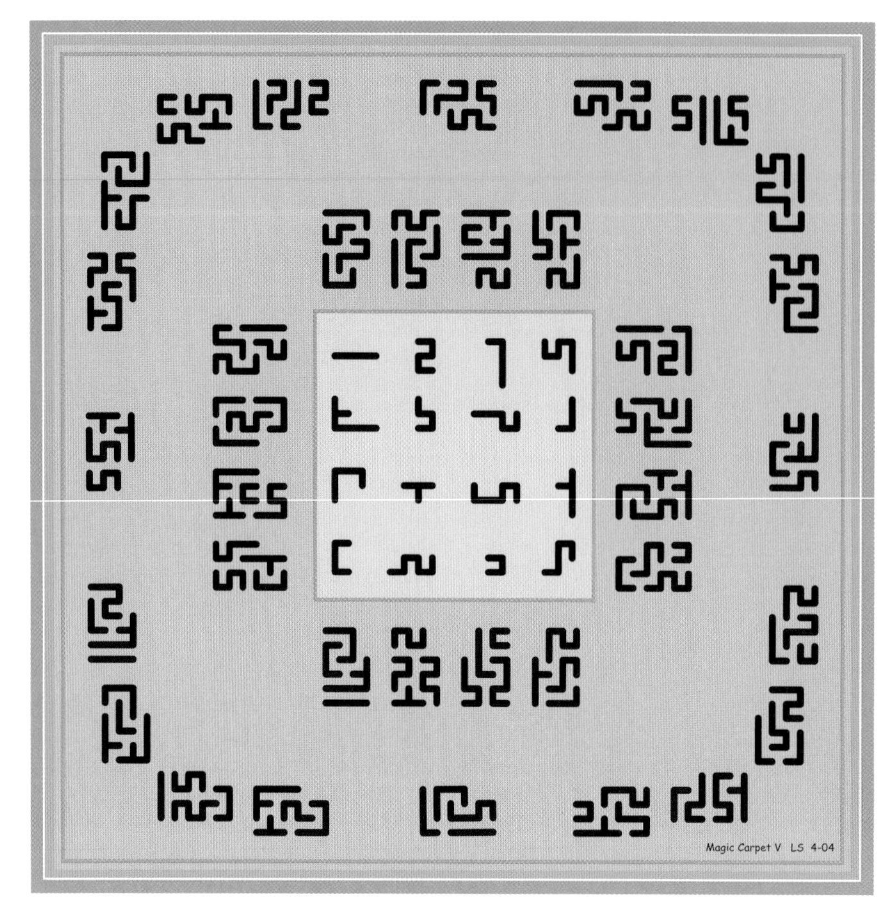

Fig. 16.7 'Magic Carpet V,' a nasik and compact square.

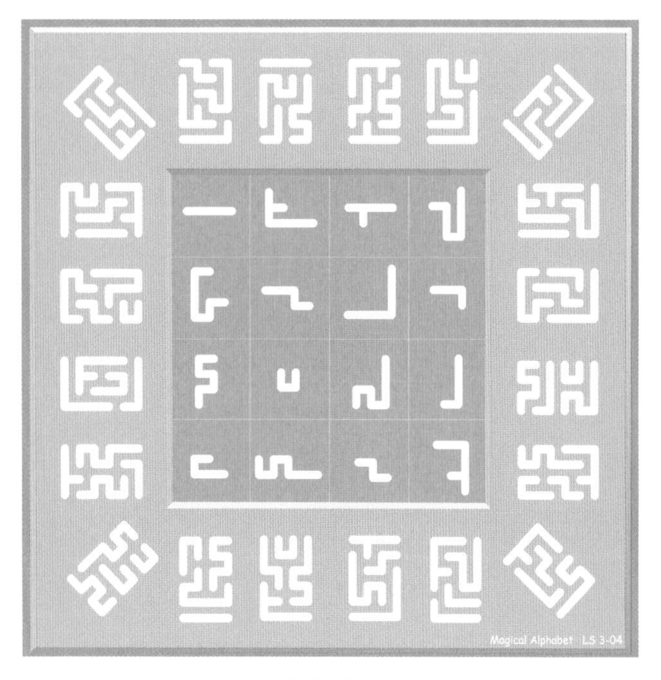

Fig. 16.8 'Magical Alphabet,' a secret message in an exotic script?

17 Form and Emptiness

Consider Figure 17.1, which is perhaps the most remarkable geomagic square we have looked at thus far.

The peculiar properties of this specimen will repay careful scrutiny. Some readers may think there is a mistake. But no, be assured that this is indeed the most extraordinary geometric magic square we have yet seen. If you have looked really attentively you may have

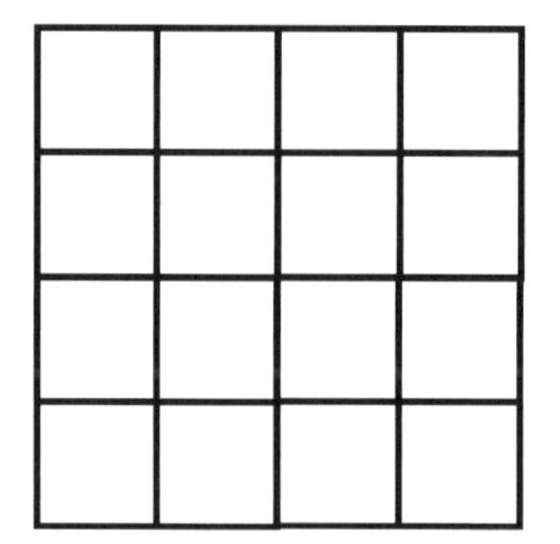

Fig. 17.1 The most remarkable geomagic square thus far seen?

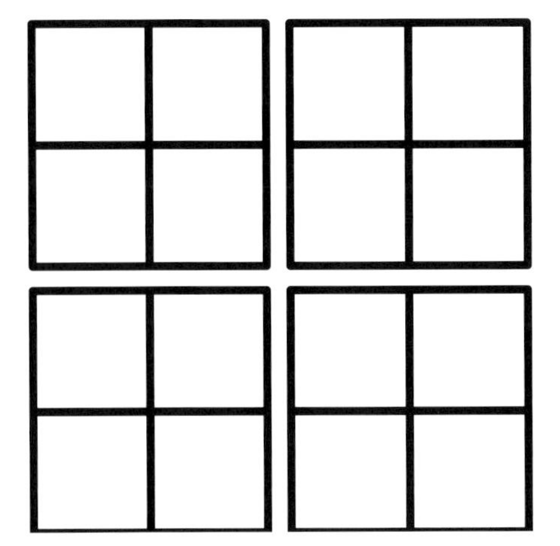

Fig. 17.2 Separating the 4 slightly overlapping sub-squares.

noticed that the figure is, in fact, built up from 4 slightly overlapping sub-squares, shown slightly separated in Figure 17.2.

Moreover, these 4 component squares are themselves composed of 4 separate pieces placed immediately adjacent to each other, as a further exploded view makes plain; see Figure 17.3.

There are 16 separate pieces here. It will be easier to understand their properties by giving each one its own color and by indicating their sizes; see Figure 17.4:

Note how every piece can be associated with a unique cell in the 4×4 array that formed our starting point. But this means that the 4 pieces belonging to any given row, column, or diagonal can be identified without ambiguity. Figure 17.5 reveals why we might wish to do so.

The 16 pieces form a self-interlocking geomagic square. The 20 targets shown look like window frames. Hence, not only can the entire set of pieces be squeezed together to produce a single square, the peculiar shape of the pieces then results in an empty framework of size 4×4. Figure 17.6 shows another view of the same structure in which a labelled card is attached to each piece.

An imploded version of the same square is seen in Figure 17.7. In the centre square the labels have been removed and the 4 quadrants overlapped to reduce the thickened central cross. The removal of the black piece

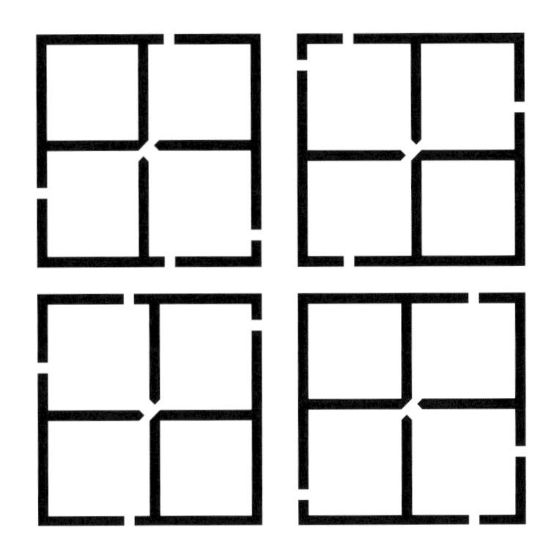

Fig. 17.3 An exploded view brings the 16 components into view.

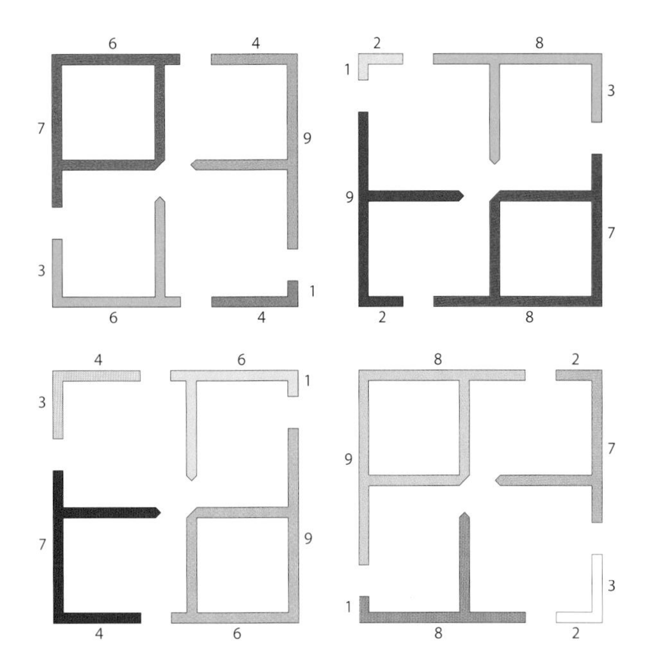

Fig. 17.4 Adding colour and size details illuminates still further.

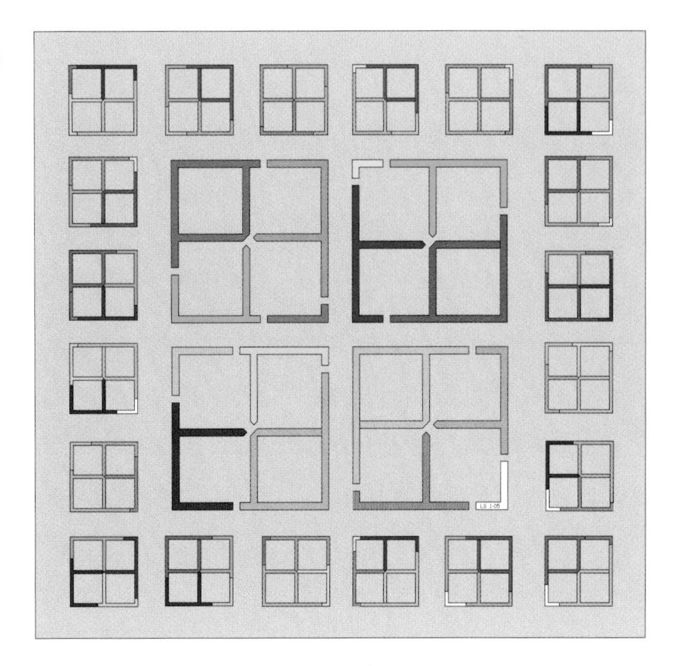

Fig. 17.5 A self-interlocking square of 4×4.

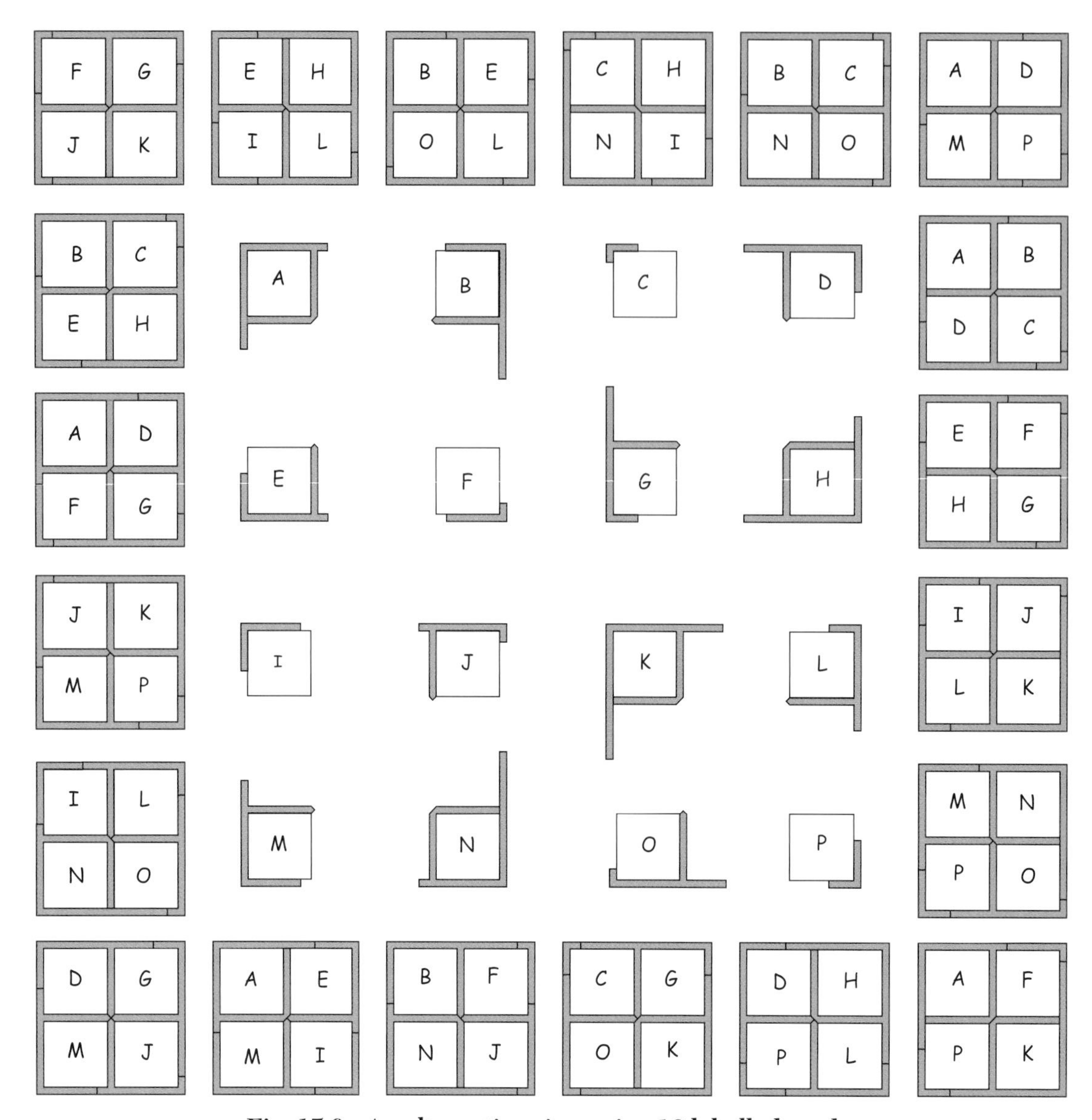

Fig. 17.6 An alternative view using 16 labelled cards.

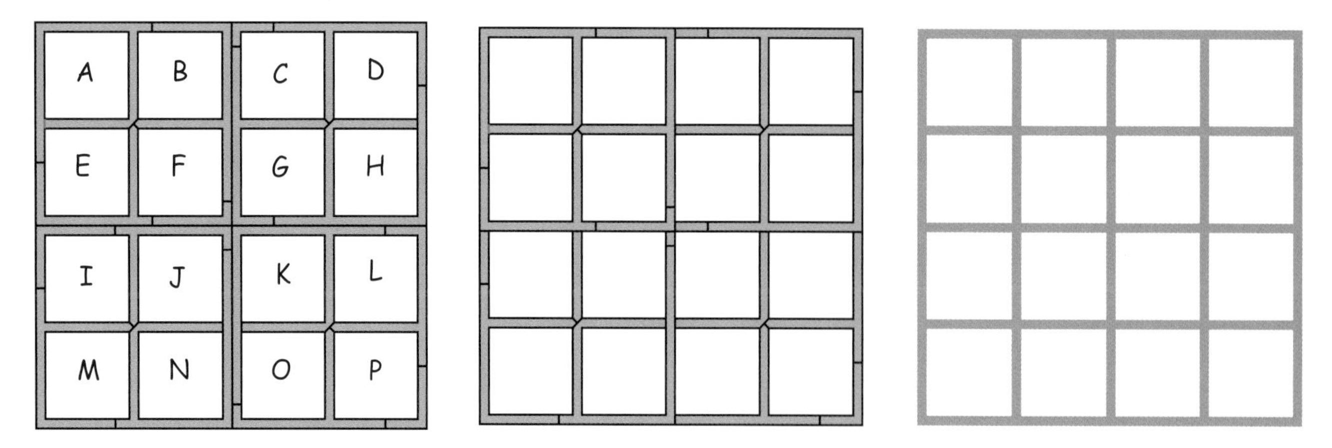

Fig. 17.7 Back to our starting point: a geomagic square lurks unseen in every 4×4 array.

outlines returns us to our starting point: an empty 4×4 array that is really a geomagic square, however difficult to recognize as such at first sight.

18 Further Variations

Excepting Figures 16.7 and 16.8, every geomagic square looked at in the preceeding two sections has been an alternative expression of the Type V template in Figure 16.1. It makes little difference which of these we may like to think of as the 'basic' square, but Figure 16.1 is a convenient choice in that its keys and keyholes are clearly distinguishable as such, which is not the case in most of the examples. Continuing in the same vein, 'Mozaic Law' in Figure 18.1 shows a further variation. Again, 20 of the 44 possible targets are shown.

Here the lengths of the L-shapes forming the right-angled corners of each piece will be found to match the corners of the pieces in Figure 17.4. But whereas Figure 17.4 displays pieces whose corners are composed of segments belonging to the *periphery* of the target square, here we are looking at their corresponding *radial segments*.

Figure 18.2 shows another version of Figure 16.1, again using polyominoes, but now different in shape to those of

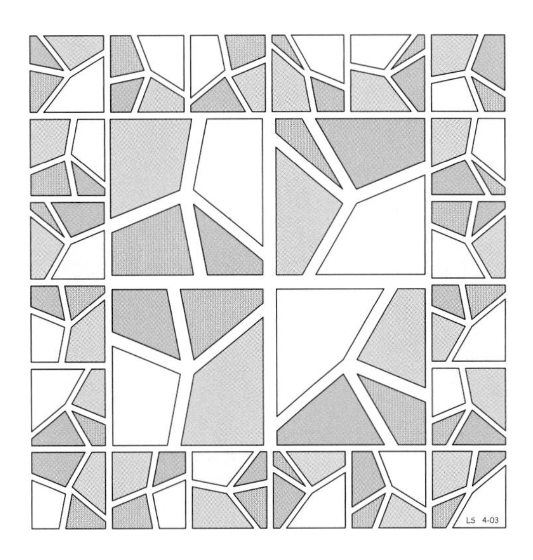

Fig. 18.1 'Mozaic Law' or the rules governing quasi-paving.

Figure 16.3. The addition of white paths to the pieces is purely for artistic effect.

Figure 18.3 explains the design of piece shapes in Figures 16.3 and 18.2. Recall that in both cases, the substrate from which we start is simply an identical square-shaped piece in every cell. Those used in Figure

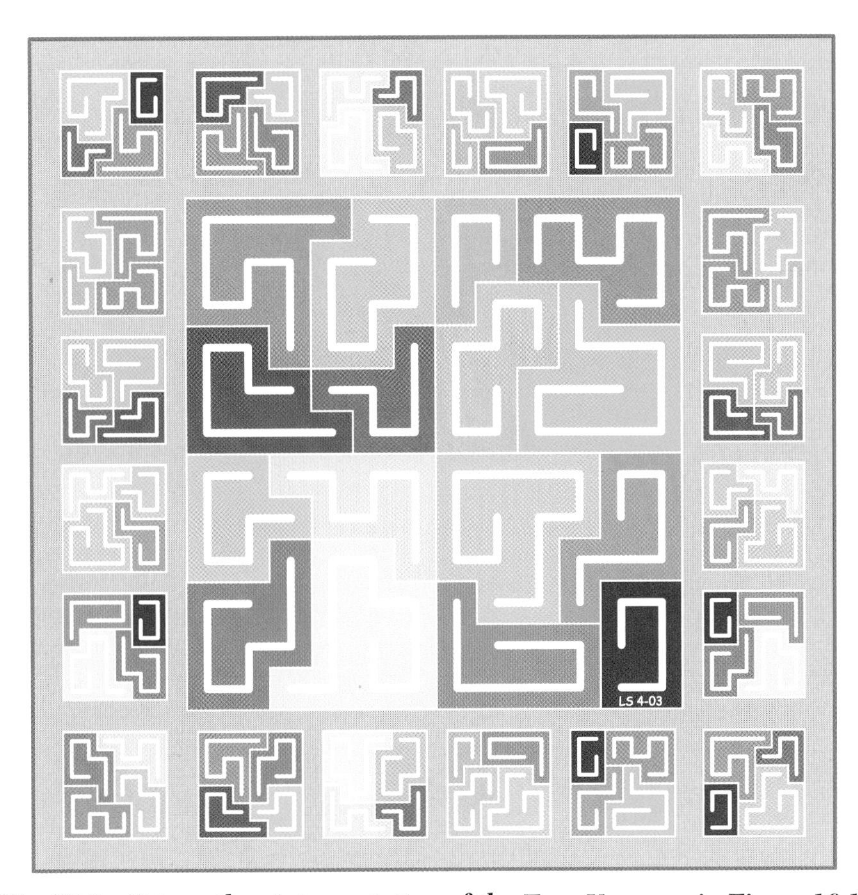

Fig. 18.2 Yet another interpretation of the Type V square in Figure 16.1.

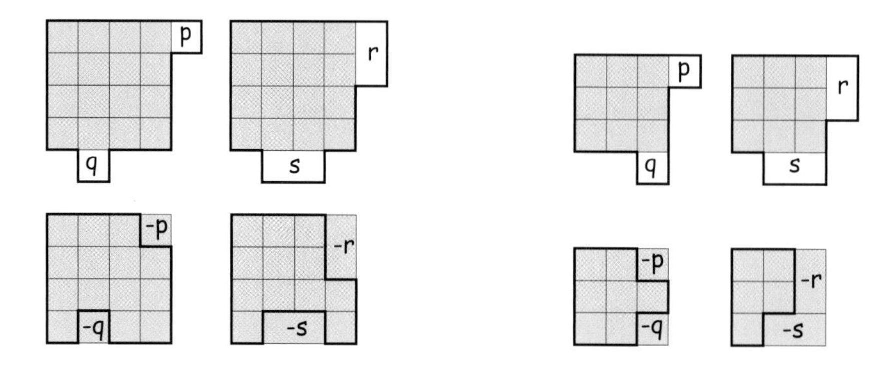

Fig. 18.3 Piece shape design in Figure 16.3 (left) and Figure 18.2 (right).

Fig. 18.4 Another Type V square. Note the absence of forked paths.

16.3 are of size 4×4, those in Figure 18.2 of size 3×3; both of which are shown shaded in Figure 18.3. Keys and keyholes are then appended to or excised from each piece as dictated by the Type V template in Figure 16.1. The key corresponding to both p and q is a monomino that is appended to the square in two distinct positions. Similarly, the key corresponding to both r and s is a domino, again affixed in two distinct positions. The 4 pieces shown in Figure 18.3 are those corresponding to $p + q$, $-p - q$, $r + s$ and $-r - s$ in the Type V template.

The choice of initial square size, as well as the shape, size, and position of keys and keyholes has to be carefully considered if repeated piece shapes are to be avoided. Trial and error aided by experience were my guides in arriving at the shapes here seen. The pieces used in Figure 18.3 are, in fact, the smallest useable polyominoes, a property that is reflected in the absence of forks when the same pieces are represented by paths, as in Figure 18.4. Here the design of the targets has affinities with maker's seals found on Chinese silk paintings, as well as Turkish Kufic calligraphy.

In the foregoing I described how a computer program that examined in turn every possible quad of 4 terms occurring in the Type I formula (the Class 1 entries) finds 52 that sum to zero (or $4A$ when A is added to each cell). Fifty-two is thus the maximum possible number of target-tiling quads that are possible in a 4×4 geomagic square of Type I. In the nasik geomagic square of Figure 10.4, every one of these 52 quads corresponds to a target-tiling piece set. Eight 'forbidden' sets reduce this total to 44 for the squares of Figure 15.4 that use an entirely different set of keys and keyholes. Now in the latter cases the variables p, q, r, and s are represented by distinct key and keyhole shapes, such that p can marry only with $-p$, q only with $-q$, and so on. But as Figure 18.3 shows, this need not always be the case. Identically shaped keys can yet be distinct in virtue of their different *positions* on the piece they modify. The argument that 44 is necessarily the maximum number of target-tiling piece sets possible is thus not automatically applicable to other cases. In point of fact the pieces in Figure 16.3 do indeed yield 44 targets, while those in Figure 18.2 yield 47. Once again, I am indebted to a computer program of Pat Hamlyn's for ascertaining these results.

Forty-seven targets can be formed with the pieces in Figure 18.2, but only 20 of them are shown in the Figure. The explanation is artistic: 20 targets fit nicely around the central square; 47 would not. Only one more however will make 48, which is exactly the right number to complete a *double* border, a goal that is realized in Figure 18.5, which is a geomagic square arrived at through experimenting with piece shapes derived from a different template.

The self-interlocking squares seen above have all been of Type V. Figure 18.5 is of Type VI, based on the template in Figure 18.6. Letters added to the pieces make it easy to identify the composition of each target. Forty-four of the 48 targets correspond to the same quads of zero-totalling algebraic terms found in previous squares. For example, the 4 pieces in the top row are a, b, c, and d.

The sum of their areas is thus $(A + p + q) + (A - p + q) + (A + p - q) + (A - p - q) = 4A$, the target area; see Figure 18.6. The 4 extra or abberent targets (over and above the usual 44) occupy the outer corners of the border: *defm*, *fgmp*, *ehmp*, and *acik*. The area of the first is thus $(A - p - q) + (A - q - r) + (A - q + r) + (A + r + s) = 4A - p - 3q + r + s$. The key/keyhole sizes in Figure 18.5 are however the same as in Figure 18.3 in which $p = q = 1$ and $r = s = 2$, so that $(-p - 3q + r + s) = 0$, as required.

In the section to follow we shall take a closer look at the special topic of normal squares, but for now I conclude this discussion of 4×4 geomagics with an alternative rendering of the previous example. Figure 18.7 shows the same 16 pieces together with their 48

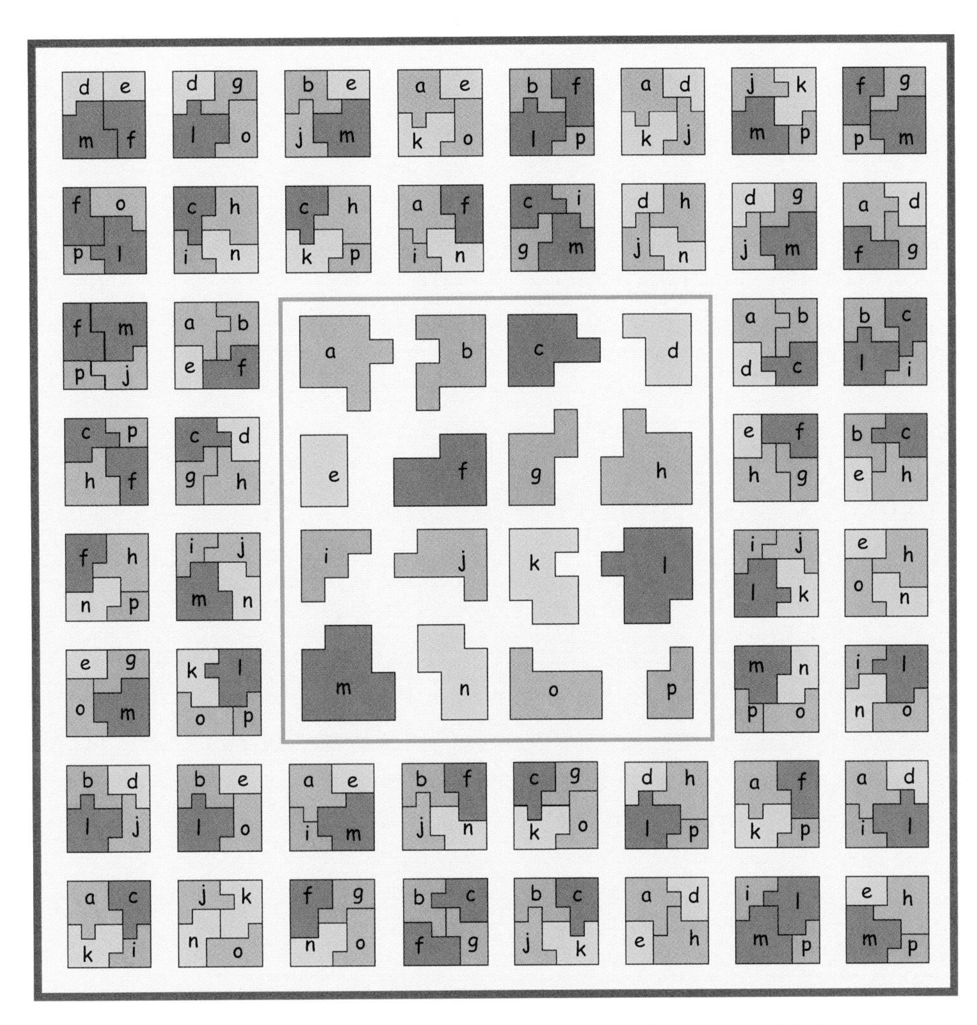

Fig. 18.5 A Type VI yielding 48 targets, enough for a double border.

$A+p+q$	$A-p+q$	$A+p-q$	$A-p-q$
$A-q-r$	$A-q+r$	$A+q-r$	$A+q+r$
$A-p-s$	$A+p-s$	$A-p+s$	$A+p+s$
$A+r+s$	$A-r+s$	$A+r-s$	$A-r-s$

Fig. 18.6 A Type VI template.

targets, but here drawn together into a single seamless square 'carpet.' Astrological signs, appropriate in view of their magical connotations, have replaced letters. I like to think that such a carpet would indeed fly, given only a carpet maker with the skill to weave it. In any case, many hours went into the creation of this design. Perhaps those who can decipher the heiroglyphics will be able to discern here my personal offering to the god of magic squares.

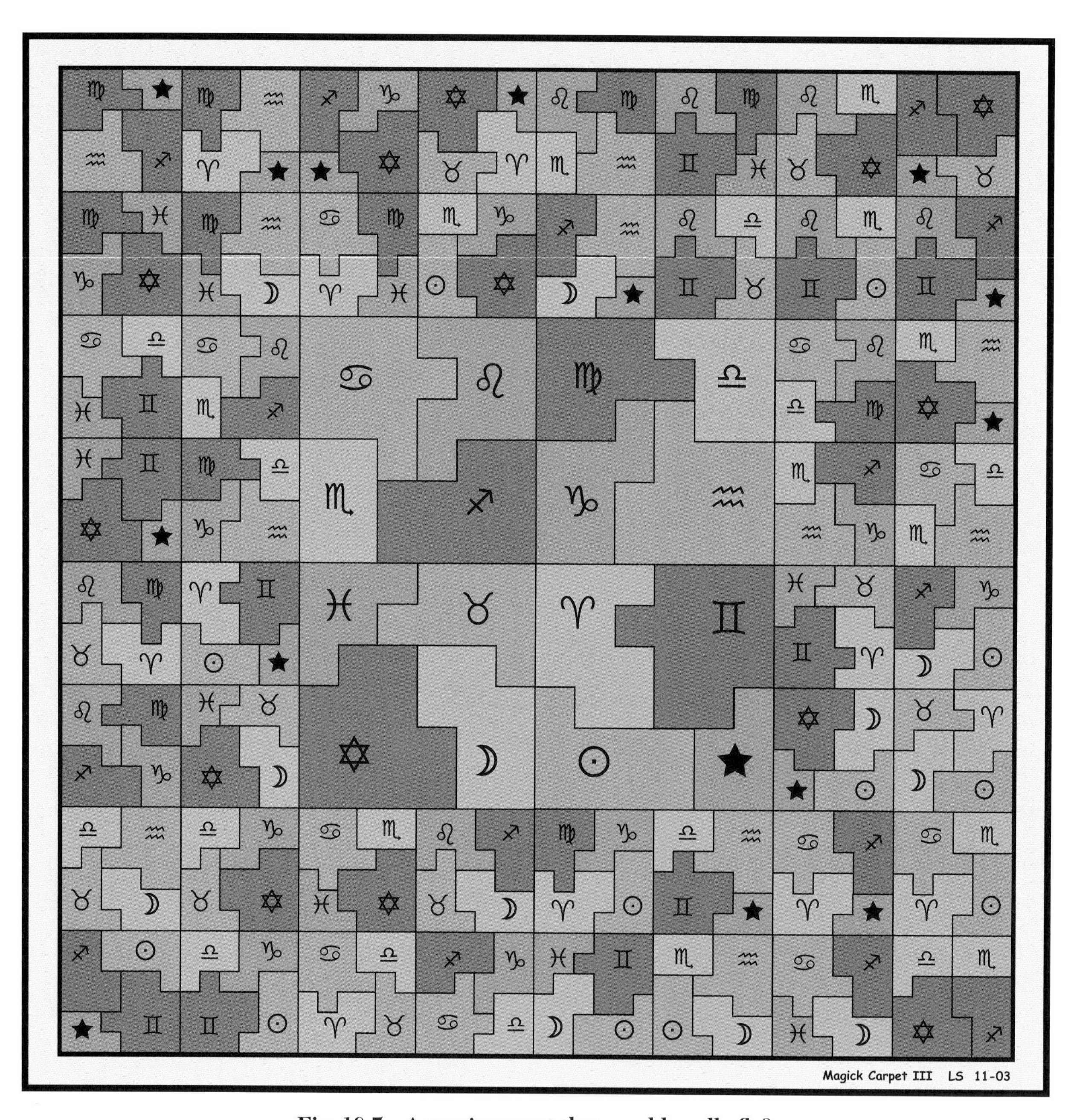

Fig. 18.7 A magic carpet that would really fly?

Part III
Special Categories

An act of magic consists in doing what others believe is impossible.

19 2×2 Squares

If this were a book on numerical magic squares then the fact that we started with squares of size 3×3 would make sense because they are the smallest possible. For a 2×2 numagic square to be magic its four numbers must all be the same, as a moment's reflection will show. But what about 2-*D* geometric magic squares; can an example be produced? I was so inured to the notion of 3×3 squares as smallest that it was some time before this question so much as ocurred to mind. When it did, however, I found myself confronting a problem that, for all its seeming simplicity, would resist every attempt at solution, even following some years of thought. Indeed, to this day, the question of whether or not there exists such a square using *connected* pieces remains unanswered. Moreover, it wasn't until quite recently – during the writing of this book – that Frank Tinkelenberg, a Dutch software developer in Leyden, came up with the very first specimen. It is one among four examples that employ disconnected pieces.

Before coming to Tinkelenberg's solution, we may note that 2×2 'semi-magic' or 'orthomagic' squares using connected pieces do indeed exist. These are squares that are magic on rows and columns only. Figure 19.1 shows an example using polyominoes of two sizes, 11 and 12.

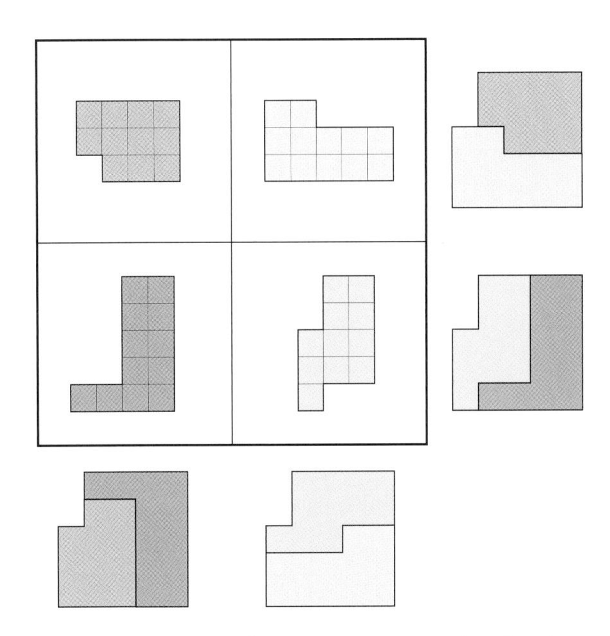

Fig. 19.1 An orthomagic square of order-2.

Clearly a fully magic specimen would have to exhibit four pieces of equal size. Semi-magics with this latter property can be found. Figure 19.2 shows a second semi-magic square, the rows and columns of which remain magic whatever the width of rectangle *R*. Note that the areas of the two green pieces are both 24, while the areas of the remaining two pieces are 14 added to the area of *R*. Hence the areas of all 4 pieces will be equal when the area of R is 10, which is to say, when its width is 5/3. However, this brings us no nearer to our goal of a fully magic specimen.

Fig. 19.2 The areas of the pieces are equal when $R = 6 \times 5/3$.

More impressive is an almost-magic 2×2 square due to Michael Reid that he produced (without quite realizing it) in response to a question I posted on an Internet forum devoted to polyforms: "Can a set of four distinct planar forms be found such that any two of them will tile the same region?" Reid's outlandish target shape and complicated 12-polyhex pieces seen in Figure 19.3 are a tribute to his mastery in the field of polyforms.

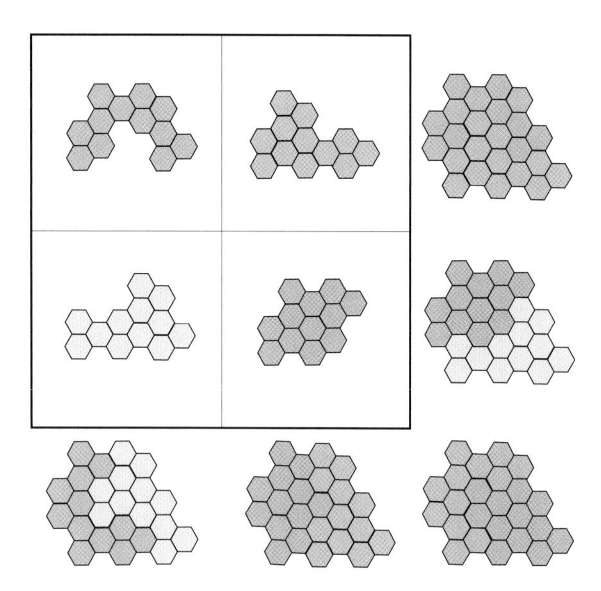

Fig. 19.3 Michael Reid's almost-magic square of order-2.

After some study, I was interested to note that the pieces in this square can be 'deflated' by performing the replacement shown in Figure 19.4.

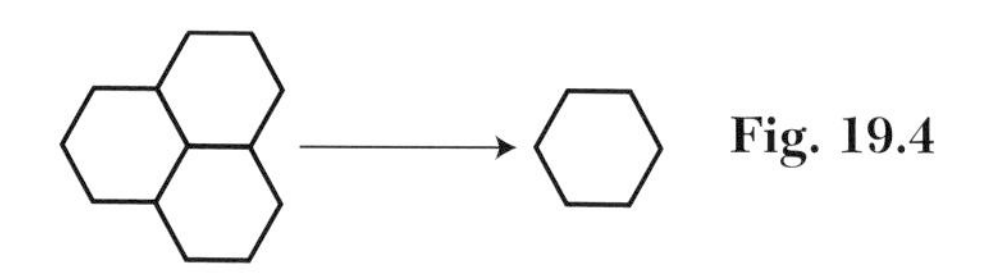

Fig. 19.4

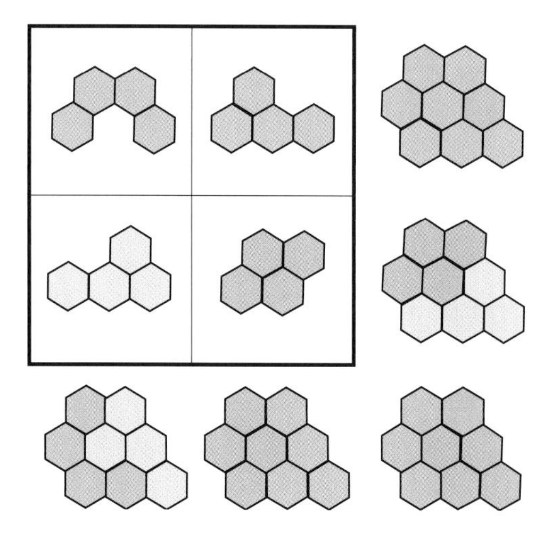

Fig. 19.5 Reid's square simplified, but one piece repeats.

The result, seen in Figure 19.5, is a simpler almost-magic square with a bilaterally symmetrical target, but unfortunately trivial, because one piece has become the reflection of another. On the other hand, regarded as a set of three distinct pieces, any two of which will tile the target, the same 4 polyhexes can be used to construct a peculiarly simple *geomagic triangle*, as demonstrated in Figure 19.6. Two similar magic triangles can be constructed using three of Reid's original pieces.

Yet a further example of Reid's inventive mind is his discovery of a non-trivial 2×2 geomagic square using three-dimensional pieces. The idea can be pictured as follows. Imagine four identical hollow hemispheres each pierced by a small circular hole in a different position. These are the four distinct pieces to be used in the geomagic square. Now take any two hemispheres from this set of four. They can be brought together and rotated with respect to each other so as to form the target, which is a hollow sphere having two holes separated by a certain fixed distance. That is all. It doesn't matter which pair of hemispheres we begin with because a suitable rotation will always be able to adjust the hole-to-hole distance to that required, whereas the lack of any reference point on the spherical surface makes it impossible to speak meaningfully about the 'position' of any such pair. (This is one of those ideas that is so simple that it has to be thought through three times before it can be understood). There are some limits to the minimum and maximum possible distances between the two holes, both of which are functions of the angles between the radii to the centre of the holes and the plane of the 'equator' joining the hemispheres.

It is interesting to note that, as with any fully magic 2×2 geomagic square, the pieces used in Reid's 3-*D* solution can be rearranged to yield three distinct 2×2 squares, all of them nasik, as shown in Figure 19.7.

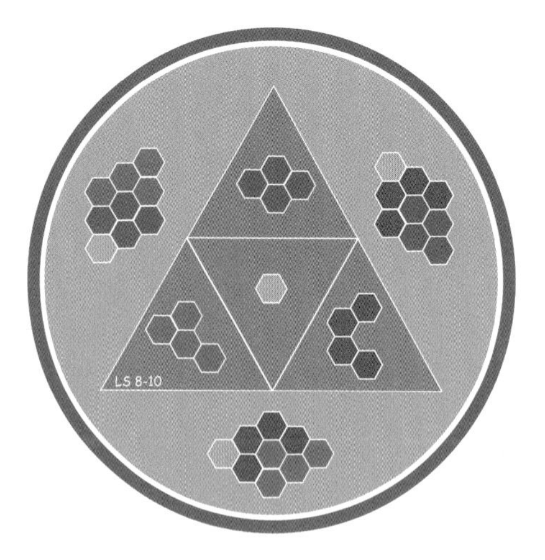

Fig. 19.6 A geomagic triangle.

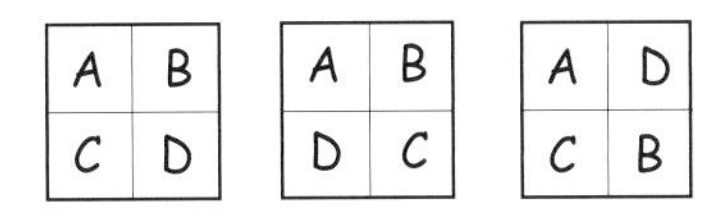

Fig. 19.7 Three distinct nasik squares.

The latter feature is borne out by the fact that on tiling the plane with any of these squares, we discover a (nasik) geomagic square in *every* 2×2 subsquare.

These were interesting advances, but it seemed to me that Reid's almost-magic square only sharpened the question: does a 2×2 geomagic square using four distinct *planar* pieces exist or not? At length, the appearance of Frank Tinkelenberg's solution using disconnected pieces was to settle the matter for good. His result, using four pieces composed of six distinct 24° segments belonging to a regular 15-gon, is shown in Figure 19.8. It is one of four variations that can be worked on the same principle. Note the two gaps in each of the targets, all six of which are drawn in the same orientation, leading to slightly rotated versions of their constituent pieces as compared with the positions of the latter in their cells. Figure 19.9 shows an alternative illustration that is possibly slightly easier to follow. In the main diagonal (\) target, the red piece appears reflected, the only such case to be found.

For me, a striking feature of this solution is the use of both disconnected pieces as well as disconnected target. As we saw in the case of 3×3 nasik squares, solutions using disconnected pieces can often be found when whole piece solutions are lacking. This is because solid pieces tile only by abutting, whereas disconnected pieces are also able to overlap each other and thus offer a larger range of tiling combinations. It is a tribute to Frank's ingenuity that he extended this flexibility-winning principle to the target itself and thus finally put to rest what had become a major open question. But of course, the Great Unanswered Question that remains is still: Does there exist a 2-*D* geomagic square of order-2 using connected pieces?

20 Picture-Preserving Geomagics

A remarkable new development that took me completely by surprise was the discovery of what I now call 'picture-preserving' geomagic squares, a wonderful idea due to Robert Fathauer, presently living in Arizona. Robert, who introduced himself to me following a talk I gave on geometric magic squares at the 2006 Gathering for Gardner 7 conference in Atlanta, Georgia, runs a firm specializing in the design and manufacture of educational aids used in teaching mathematics. As such, geomagic squares held for him a peculiar interest, his idea being to produce physical models using real pieces that might be used to illustrate elementary theorems or other aspects of number theory. Initially I was a little unclear about these proposals until he sent me the square shown in Figure

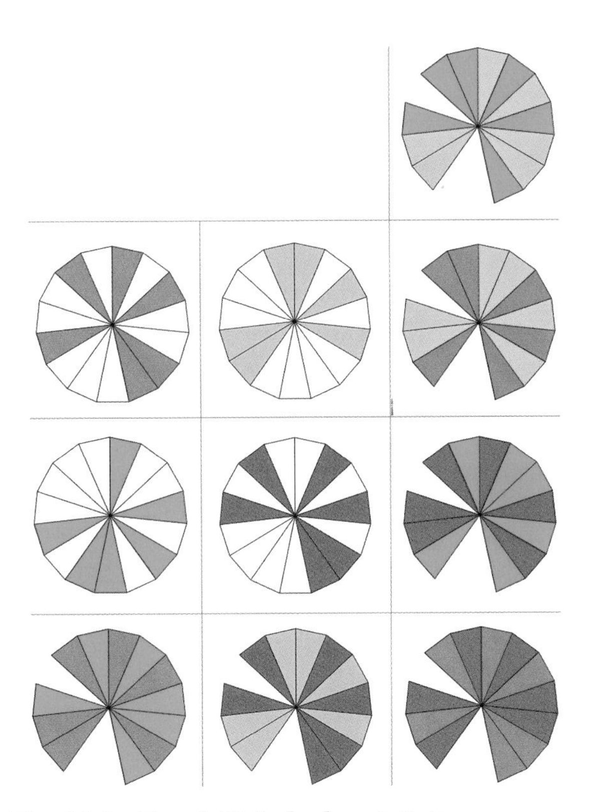

Fig. 19.8 Frank Tinkelenberg's 2×2 square.

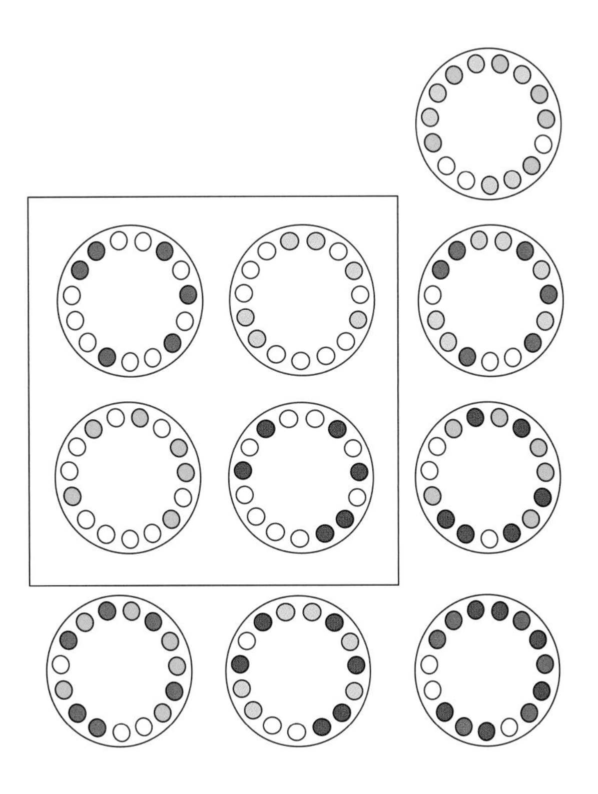

Fig. 19.9 An alternative version of Tinkelenberg's square.

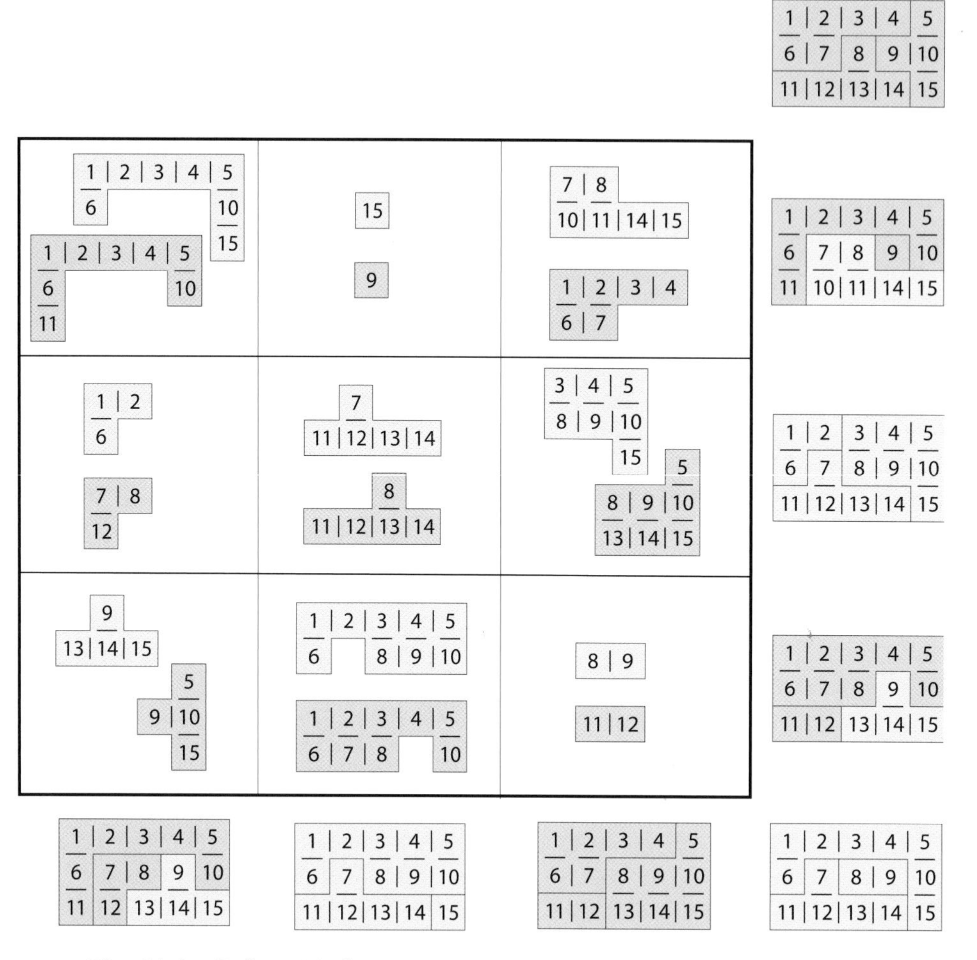

Fig. 20.1 Robert Fathauer's picture-preserving geomagic square.

20.1. It is a modified version of Figure 6.1 on page 12, which I had sent him previously, being one of the 1,411 normal geomagic squares tiling a 3×5 rectangular target. The square can be seen again in Figure 20.3 (left).

At first sight it may seem that there are two pieces occupying each cell in Figure 20.1. This, however, is not really the case, the intention being rather to show what is printed on the front and back faces of the single piece occupying each cell. The two shapes in each cell are thus reflections of each other, although in several cases the flipped twin has been rotated so as to bring the numbers printed on both sides right side up. Simply flipping a piece about a vertical or horizontal axis, without rotation, will often reveal numbers that are printed either upside down or sideways. To avoid this, the orientation of each piece shown in the cells has been chosen so as to show the digits right side up, which is the same as that in the targets in which they appear.

Two distinct shades of blue distinguish fronts from backs, making it easier to see how targets are composed of flipped and/or unflipped pieces. In a physical realization using real pieces this would be unnecessary. Using the same tint on both sides would then make for targets of uniform color. As in the square on which it is based (i.e. Figure 20.3), the target is again a rectangle, but now bearing the numbers from 1 to 15 in serial order. The idea was to use this as a didactic device in combination with a worksheet discussing how to sum the first n natural numbers. It took me a few minutes to figure out the extraordinary implication of what Fathauer had achieved in his modified version of my square.

Every piece bears numbers, front and back, such that the target is not merely a rectangle, but one in which these numbers appear in serial order from left to right. The individual numbered cells thus appear in exactly the same position in every target, and the same would be true whether or not these cells bore numbers or some other graphical device. In particular, that device might be a square fragment belonging to a given position within *any* 3×5 field or 'picture.' But in that case the pieces in the cells would themselves correspond to larger picture fragments that assemble in the targets so as to reproduce the entire 3×5 field. Targets will then not only share the same outline shape, their surface detail will also be identical, point for point, or if you will, pixel for pixel. Hence the appellation 'picture-preserving' (or 'pp') for a geomagic square of this kind, although in the case of Fathauer's square the 'picture' is less a tableau or view than it is an austere table of numbers. Figure 20.2 exploits the pp-property to better effect, the target here being the Five of Hearts playing card. To avoid overcrowding, the reverse sides of the pieces are not shown in the cells, but can be inferred from the targets. For instance, from the left-hand-column target can be seen that the square piece that is blank (seen in the left-hand column centre cell) bears two half-hearts on its reverse side.

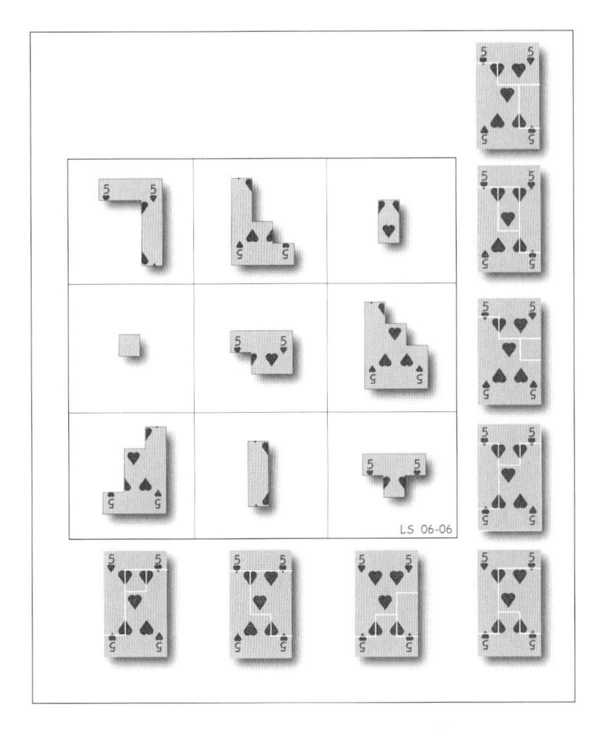

Fig. 20.2 A picture-preserving square.

What is the essential property that distinguishes picture-preserving geomagics from others? It is that wherever any given piece occurs in different targets, it always occupies exactly the same position within each of them. For only thus will every picture fragment always find itself in its appropriate place within the assembled picture. Yet, since the flip side of a reflected piece may bear distinct markings, in this respect a reversed piece may be treated as a separate case. These points will becime clearer on examining the differences between Fathauer's square and the square from which it was derived, shown in Figure 20.3. For example, the grey octomino in the top left hand cell of the latter does indeed occupy the same position in all three targets in which it appears. But the blue hexomino in the top right hand cell occupies distinct positions in different targets, a feature that will disrupt picture preservation. A reflected version of this same piece occurs just once, in the top row target only, which is unproblematical, because the image depicted on its back side can be adjusted to suit that distinct region of the picture it occupies. Observe also that the domino in the bottom right hand cell may or may not be said to appear in three different targets, depending upon whether one or more of these is regarded as flipped or not. In any case, it was Fathauer's insight to see that with a few judicious adjustments to the orientation of certain targets, these problems could all be overcome, as witnessed in his modified version of the same geomagic square seen in Figure 20.3 (right).

Here the previously troublesome hexomino now occupies the same position within two targets, as does the domino, whose seeming third appearance in the diagonal target can now be interpreted as a *reflected* domino, occurring but once. Similar comments apply to the tetromino in the bottom left-hand cell, which, like the monomino, domino, and triomino, is another case of a piece

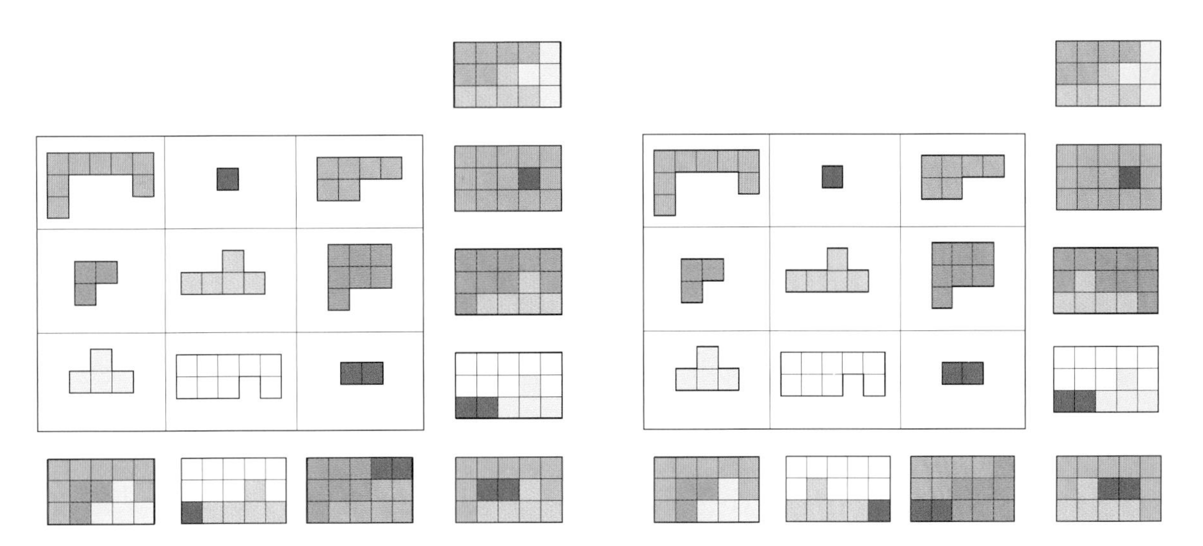

Fig. 20.3 Converting an ordinary square into a picture-preserver.

whose flipped and unflipped forms are congruent. Figure 20.3 (right) is thus suitable for use as a pp-geomagic square, as indeed it has been in providing the basis for Figure 20.1.

Having grasped the principle underlying Fathauer's square, it was natural to start re-examining existing geomagics in the hope of discovering still more picture-preservers. Was it significant that his square is derived from a specimen using polyominoes of *consecutive* sizes? Figure 20.2, an early find, which is also among the 1,411 geomagics using consecutively sized polyominoes, suggested it might be, although subsequent examples lent no support to this idea.

The picture-preserving property is perhaps easier to follow in Figure 20.4, in which the 'picture' is no more than a yellow circle centred on a green square. Seen to right and below are reflected versions of the eight targets (flipped about a vertical axis), revealing the jumbled patterns incidentally created on their reversed sides. This specimen is based on Figure 20.5, an earlier discovered (symmetric) geometric square. Interested readers may like to examine the small, yet subtle, changes to the targets that converted this initial find into a picture-preserver. Note that use of a circle as a target picture makes for a special case that forms an exception to the general rule enunciated previously. That the same piece must occupy the same position within different targets is no longer necessarily true, since the symmetry of the circle means that it can be repeatedly rotated through 90° without any change to the 'picture.'

The square target in Figure 20.5 is of size 6×6. An obvious idea suggested by this was to replace the circle

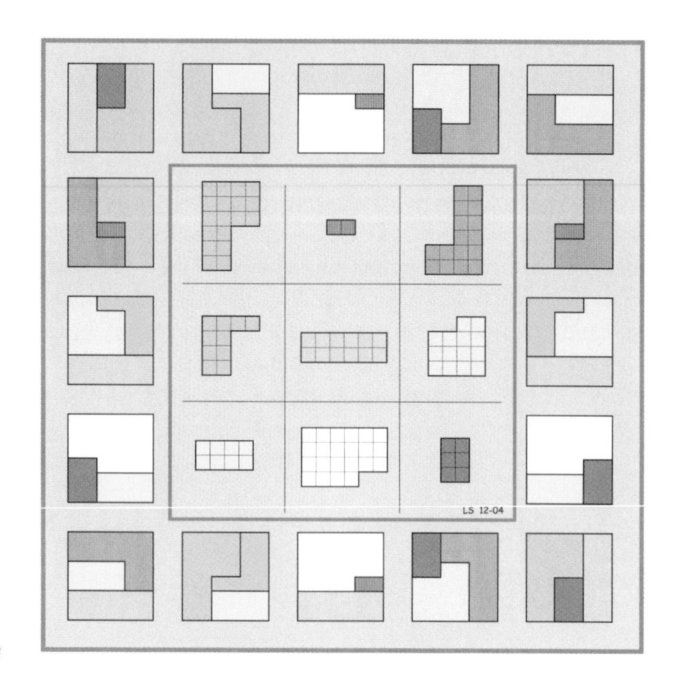

Fig. 20.5 A potentially pp-square.

motif with a picture showing a 6×6 numagic square as target. Piece edges would then correspond with grid lines on the 6×6 array, much as in Fathauer's numbered 3×5 target. An experiment showed this to be quite feasible, although the swarms of numbered cells involved make for a pictorially inelegant result. In view of this, a better idea seemed to be a target using a visually simpler square such as the *Lo shu*. Grid lines in the square will then no

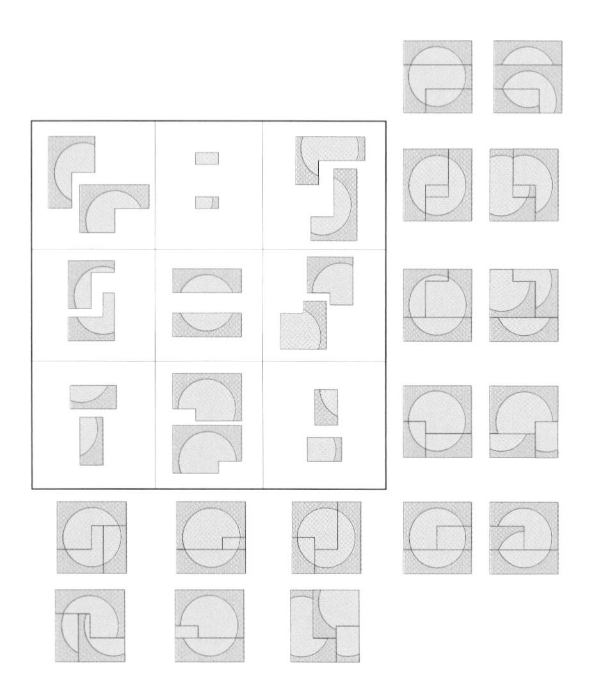

Fig. 20.4 The 'picture' is a yellow circle on a green square.

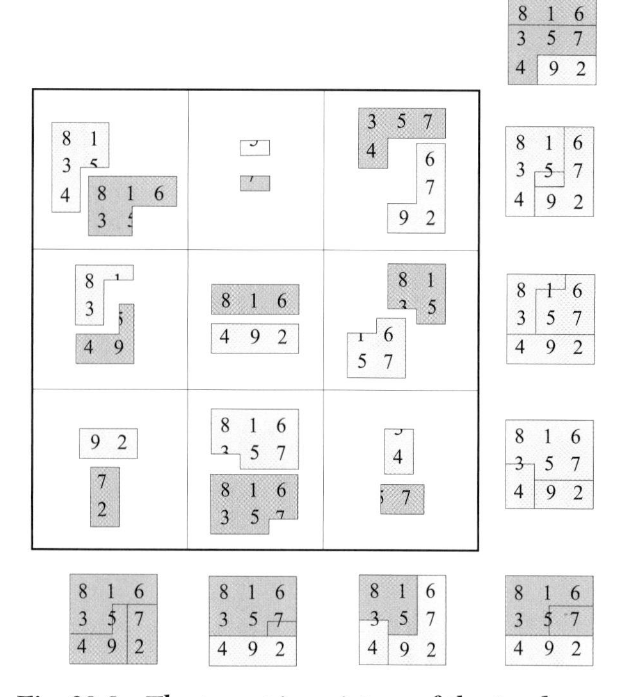

Fig. 20.6 The target is a picture of the *Lo shu*.

longer correspond with piece edges, and in fact some numerals end up becoming bisected, an outcome that, in retrospect, seems less than satisfactory. Figure 20.6 shows the resulting square. Light and dark shading distinguish piece fronts from backs.

It was subsequent to these early experiments with picture preservers that investigations in another direction brought to light a 3×3 geomagic square with some remarkable properties. Among these was the potential to be used as a picture preserver, the target being a 4×4 square with a missing cell or hole; see Figure 20.7. The hole might seem a nuisance, but with a little ingenuity could be turned to advantage in creating a kind of mathematical joke. Again, my idea was to use a traditional magic square as target picture, while the trick employed was to take a numerical magic square in which the number zero occupies the same position as the hole. In this way zero will not appear, a disappearance that leaves the arithmetic unaffected. The result can be seen in Figure 20.7. With a little patience the numbers appearing on the backs of the pieces are easily deduced from a careful examination of the targets.

I have mentioned that Figure 20.7 has remarkable properties. That it is a picture-preserving square with square target, has already been seen. That it is also a *normal* square may already have been noticed. Before resuming our discussion of pp-geomagics, a brief glance at a further characteristic of this square will prove of interest.

Different as it may appear, Figure 20.8 is in fact an alternative presentation of the square pictured in Figure 20.7. It is an alternative rendering that seeks to highlight a further peculiar attribute of the square. Here again I have used dots and lines to represent pieces, a detail of no further significance beyond indicating the square's link with the *Lo shu*. However, taking a closer look at the targets shows that they are of two kinds. To right and below are the usual row, column and diagonal targets corresponding to a 4×4 square with inner hole, exactly as in Figure 20.7. But to left and above are found eight slightly different targets in the form of a 4×4 square *without* hole. Admittedly, the eight triads of pieces that compose the latter targets are not harmoniously disposed within the square, as are those forming the rows, columns and diagonals. However, the fact that in each case they fit together to pave the same 4×4 square target is a feature that required some effort to achieve. In fact, the construction of this square arose in the wake of a simple observation concerning the numbers in the *Lo shu*. This was that in addition to eight sets of three numbers summing to 15, there exist a further eight sets of three numbers summing to 16. The notion of a geomagic square that might echo these repeated totals with repeated targets then suggested itself, with the culmination seen in Figure 20.8.

Returning now to the topic of picture-preserving squares, it is amusing to ponder that if the target picture employed can itself be a magic square, then it could even be a picture of the geomagic square that is *itself*. The result would then be a specimen in which the so-called 'Droste effect' would be prominent. Pieces would show fragments of the complete array, upon which would be seen smaller pieces nestling within their cells, upon which

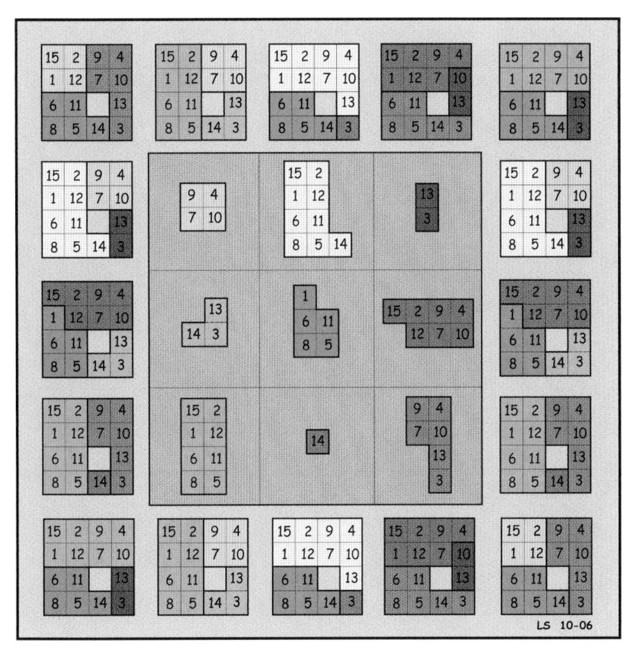

Fig. 20.7 A 3×3 geomagic square with a 4×4 numagic square as target.

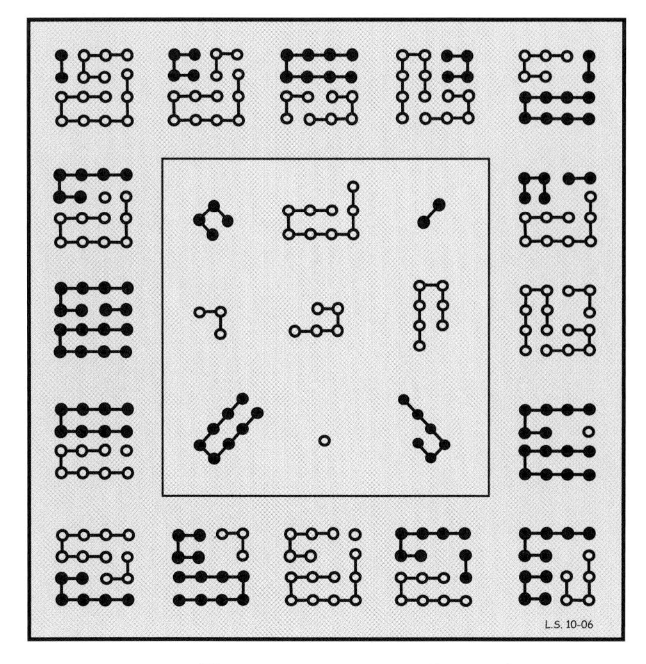

Fig. 20.8 Different as it seems, this is the same square as in Figure 20.7.

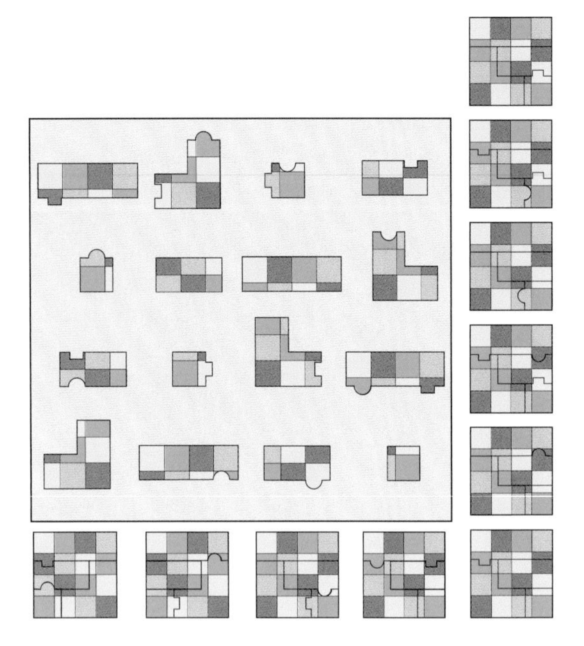

Fig. 20.9 A chequered target reveals the picture-preserving property of Figure 9.7.

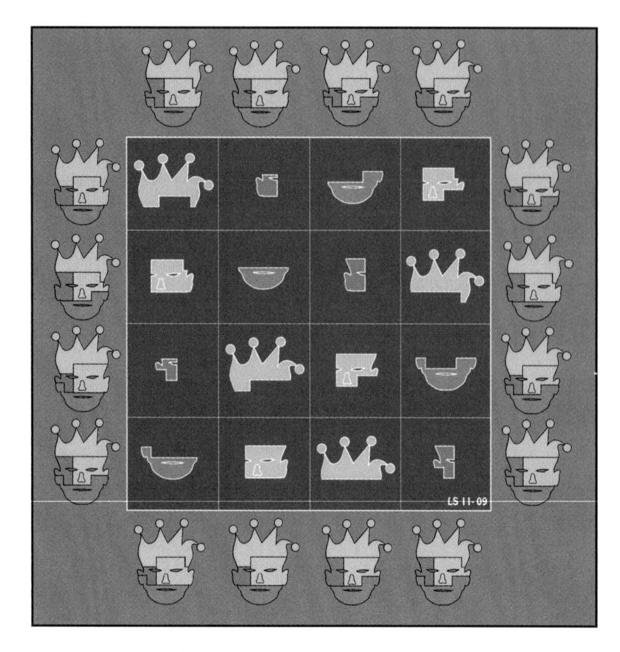

Fig. 20.10 'The Joker' that could take many guises.

could be seen still smaller pieces cradled within their cells, and so on, in endless regression. However, I suspect the result would be too visually confusing to make it worth the effort of trying to execute.

The pp-geomagics looked at thus far have all been of size 3×3. It is easy to see than no 2×2 examples will be found. Given that each piece would need to occupy an identical position in at least two different targets, the shape of the remaining target area, which is to say, the shape of the second *piece*, would then have to be the same in each case. A non-trivial 2×2 pp-geomagic square is thus impossible.

On the other hand, picture-preserving 4×4 squares can indeed be found. Figure 20.9 revisits a Latin-based square first met with in Figure 9.7. The repetitive target structure that then seemed unsatisfying is now put to use in creating a pp-geomagic square showing a square target with a simple chequered pattern. The reverse sides of the pieces can be viewed in the row targets, where they appear slightly reduced in size.

A more decorous specimen is 'The Joker' of Figure 20.10, which bears many characteristics in common with Figure 20.9 yet is distinct in design. Its template is shown in Figure 20.11, while that of Figure 20.9 appears in Figure 9.6 . As the joker's flamboyant headgear suggests, the outline of the target need not be fixed as shown, but could assume a great variety of different shapes. But at a price, to be sure, for although all four quadrants are magic, the two main diagonals are in fact not.

One more example of a square that can be used to create a pp-geomagic square of order-4 is Figure 10.4 on page 30, a latin-based example that is nasik. Piece-ordering in the targets must be changed first however, although the resulting picture-preserver will not be nasik.

There remains a further interesting property of pp-geomagics that is worth mentioning. Readers may already have noted that the picture-preserving capability extends beyond the plane to embrace the third dimension. By this, I refer to the fact that the targets in a pp-square can include pictures carved *in relief*, so long as we are prepared to ignore that part of the target structure on its reverse side, protruding backwards behind the plane that

$A+x+y$	$B-x$	$C+x$	$D-x-y$
$D-y$	C	B	$A+y$
$B-x-y$	$A+x$	$D-x$	$C+x+y$
$C+y$	D	A	$B-y$

Fig. 20.11 The Joker's template.

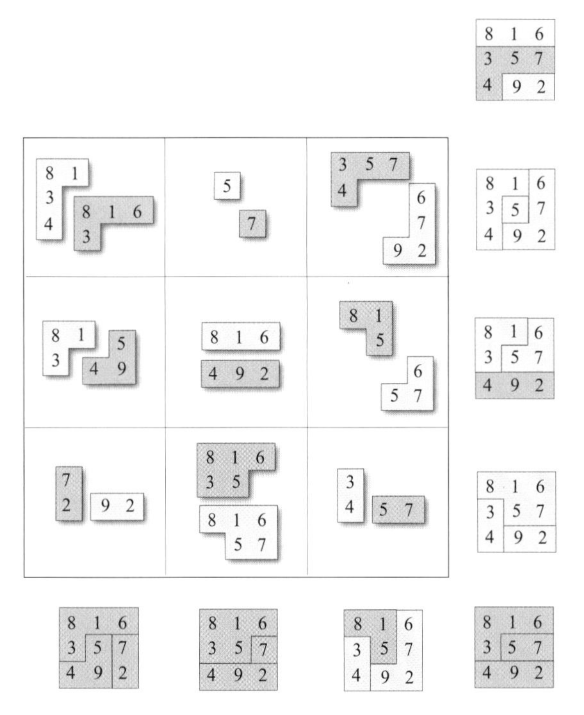

Fig. 20.12 Is this a trivial square, or not?

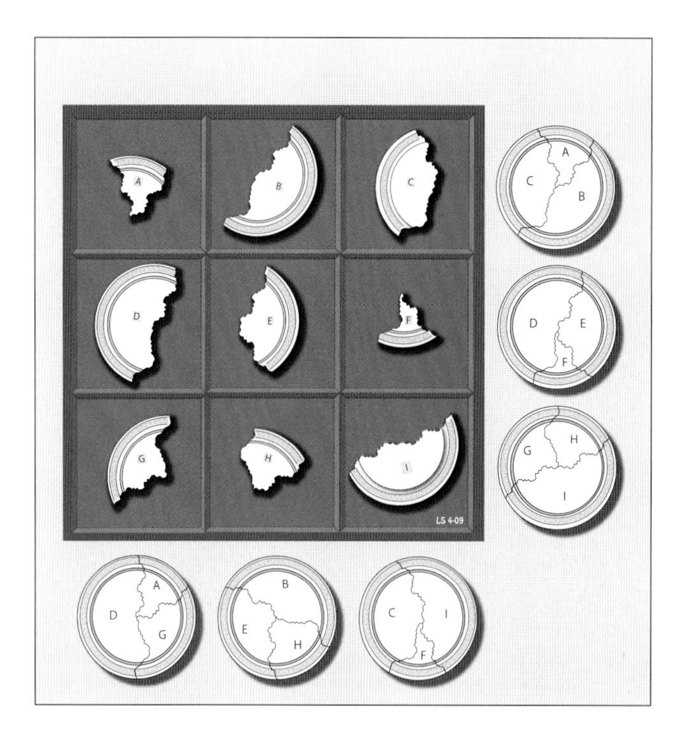

Fig. 20.13 'The Archaeologist's Nightmare.'

is its floor. To grasp this, imagine the picture in Figure 20.4 is not a circle but a solid cone, seen from above, the pointed end. The various pieces thus bear complicated dissections of this cone, which has been sliced by planes that are perpendicular to its base. On the reverse sides of pieces are similar dissections that extend away from the viewer rather than toward. In the targets we view the reconstructed solid cones, although behind them is an unholy chaos, equally three-dimensional, corresponding to the jumbled patterns seen in the reversed targets of the same figure. Granted that this feature would not seem to be of any particular 'use,' I find it intriguing. My own belief is that the last word on pp-geomagics has yet to be said.

A final point of interest is to see how the introduction of the pp-geomagic square brings with it the need for a clarification in our definition of 'trivial'. Figure 20.12, a supposedly trival square, illustrates the point, its square target, again a picture of the *Lo shu*, being formed from pieces of repeated shape, which are nevertheless distinct in virtue of their *unique marking*. So, we may ask, is Figure 20.12 really trivial or not? The answer is of little import, of course, but the point is worth bearing in mind.

I can hardly conclude this section on picture-preservers without showing an unusual find of order-3 that lent itself to a playful elaboration. There exists a special kind of picture-preserver that requires no markings on the back sides of its pieces because every target can be assembled without reversing any piece. Such a square is seen in 'The Archaeologist's Nightmare' of Figure 20.13, where the implied concavity of the plate would make flipped pieces inappropriate. Given suitable marking on the hidden sides of the present pieces, the potter's trademark could be made to appear on the underside of each reassembled plate. Although not shown, the three shards on each diagonal can also be glued together to form the plate.

It may interest readers to learn that I submitted this picture to the *British Journal of Archaeology*, confident of its being enthusiastically received as a humorous filler item for the journal. However, I was wrong. They were not amused.

21 3-Dimensional Geomagics

The allusion to solid targets in the previous section brings us nicely to the topic of 3-dimensional geomagic squares. As indicated at the outset of this essay, geomagics may use pieces of any dimension, although for practical reasons 2-*D* types are bound to exert the greater appeal, since they are both easier to produce, as well as simple to present on the page and to think about and discuss. 3-*D* squares are not merely less ubiquitous, and thus more difficult to track down, they do not lend themselves to presentation on paper, being best appreciated in the form of solid objects that can be picked up and handled. And of course, the manufacture of such material objects is a demanding enough hobby in its own right, without adding the exigencies of writing a computer program for discovering the geometrical forms that are to be constructed. Indeed, bearing in mind the sheer work involved in exploring planar types, in retrospect it seems surprising I ever started in pursuit of the far more demanding 3-D specimens. But the truth is that I was spurred on by the attractions of one particular solid target that caught my fancy.

If the aim was to construct a geomagic square using solid pieces that fit together so as to complete a solid target then a square of order-3 was the obvious case to consider. Being unable to think up any alternative approach, a computer program performing a brute-force search offered the only possibility of discovering a specimen. The nine pieces used in the square would thus be *polycubes*, which is to say, the 3-D equivalent of polyominoes, or shapes that can be formed by sticking together unit cubes, face-to-face. As previously, the number of unit cubes in the target would therefore be a multiple of 3, but it must not be so large as to make piece numbers prohibitive. The larger the piece size, the more of them there are, with their numbers growing explosively as size increases. Pieces of size 10, or decacubes, were the biggest my software could handle. After weighing things up, it became clear that the ideal target, both in practical and aesthetic terms, would be a 3×3×3 cube. It was the irresistable simplicity and symmetry of this cubic target that encouraged me to start work.

Twenty-seven cubelets in the target mean that the centre piece in the geomagic square will be of size 9, a nonacube. Given 10 as the largest permissible piece, there exist but two possibilities in terms of the distribution of remaining piece volumes, uniform and latin, as shown in Figure 21.1.

A square using nine nonacubes seemed to me the most desirable, not merely for the sake of balance, but because the chance of finding a solution yielding even

9	9	9
9	9	9
9	9	9

9	8	10
10	9	8
8	10	9

Fig. 21.1 The two possible piece volume distributions

more than the necessary eight magic triads, one for each row, column and diagonal, would be greater. As in the program discussed earlier, the search for a solution relies on a list L showing every triad of distinct nonacubes than will 'tile', or pack, the target. In the present case this entailed 159,177 entries in L, a figure that called for a more efficient algorithm than the simplistic example previously described, if running time was not to become excessive. Even then, it took well over a week for the program to examine all possibilities, none of them, as it turned out, yielding the geomagic square sought. It was a near thing though. The program found some half dozen cases of squares in which all rows and columns, plus *three* diagonals (one main, two broken) were magic. But no case in which both main diagonals were satisfied. It was odd to be frustrated by specimens that embodied even more than was asked of them, and yet had to be disqualified for not doing so in exactly the way required.

Attention thus turned to the second alternative, the new list L now detailing 916,008 triads of 8, 9, and 10-sized target-packing polycubes, which is almost six times longer than the previous list. As it happens, latin-type piece

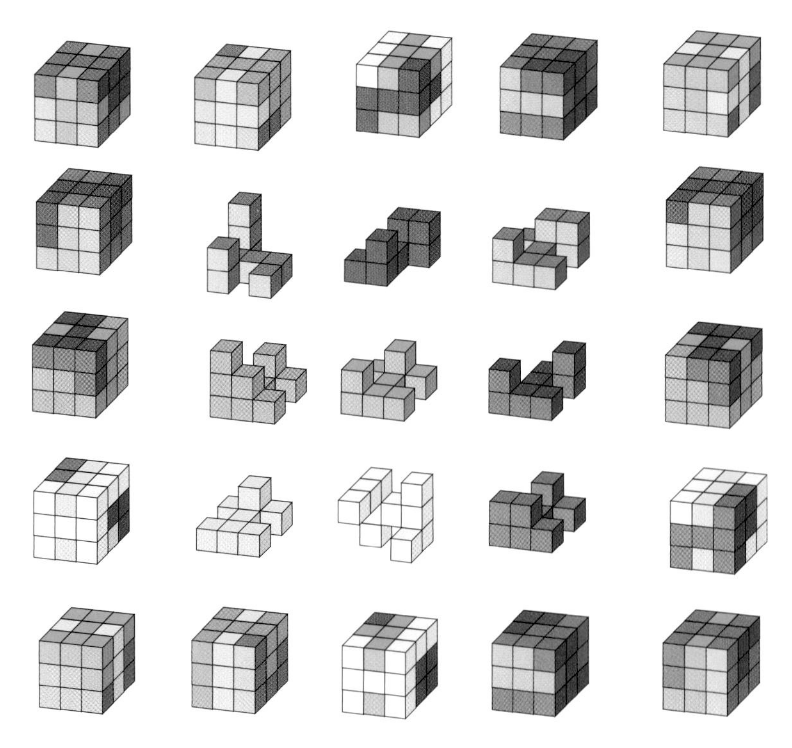

Fig. 21.2 A 3-D geomagic square with 3×3×3 cubic target.

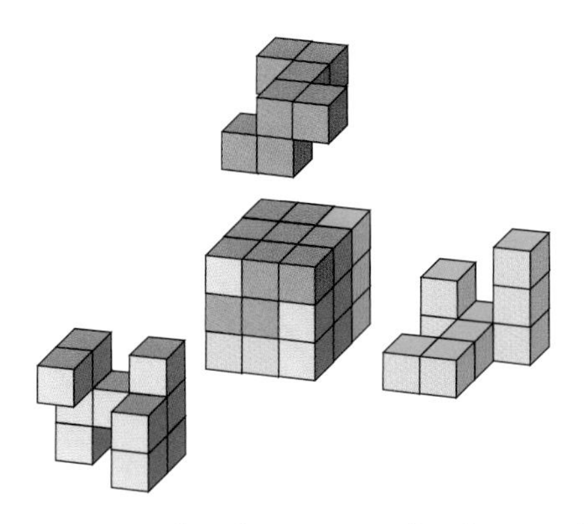

Fig. 21.3 A cube that cannot be dismantled.

distributions can be tested a good deal faster than uniform types, so that running time could remain within practical limits. And sure enough, before long a bleep from the PC indicated a solution had popped up. In concrete terms, that meant that a 3×3 array containing piece numbers had been dumped to a text file. All that was now required was to open the file, decode the numbers back into the polycubes they represented, and then check that they fitted together in the ways required. How does one do that?

My answer was already waiting in a cigar box containing a roll of double-sided sticky tape, along with dozens of pre-sawn wooden cubes measuring around half an inch on a side. This may not sound like a sophisticated method, but it is effective. A couple of hours later saw a servicable, if delicate, set of nine polycubes lined up for the crucial test. Taking up three pieces at a time, I tried fitting them together. Surprisingly, it was often far from easy. In each case it was quite a puzzle, a contingency that had not previously occurred to mind. But with a little patience the right configuration would soon be found, whereupon the three pieces slid snugly together to form the cube.

Except in one case, which defied every effort. At first this was worrying. Was there a bug in the program? But careful examination soon explained the cause, which was instructive. As in the example shown in Figure 21.3, even though they *pack* the cube, the peculiar structure of the three pieces makes their *assembly* into a cube impossible. Here the red and green pieces can be put together as required, but the blue one, which can never marry with the red, is still unable to slide into place. Or the blue and green can be put together, but the red will be left outside. In similar fashion, the fact that a cube is formed of three polycubes is no guarantee that it can then be dismantled simply by pulling and pushing. The solution discovered by the program was thus a genuine 3-*D* geomagic square, but one with a fatal flaw from the demonstrator's point of view. Here was another contingency that had not been anticipated.

Happily, the program went on to discover a wealth of further solutions, many of them unmarred by this feature, and some of them even including extra magic triads (although no demi- or semi-nasiks among them). Figure 21.2 shows a favourite specimen, wooden and metal models of which adorn my work table, even as I type. This particular instance is pleasing in realizing eight non-trivial assembly puzzles in one. I'm afraid that few visitors to

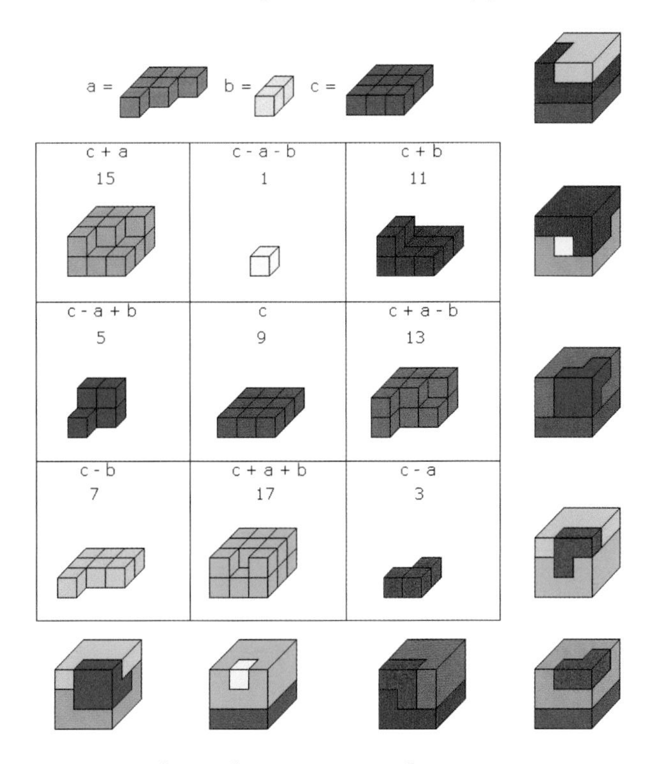

Fig. 21.4 Aad van de Wetering's elegant 3-*D* geomagic square.

the house have escaped an obligatory trial of their solving skills. To assist the viewer, the cubic targets flanking each row, column, and diagonal, are pictured as if seen from two diametrically opposed viewpoints. Nevertheless, such pictures of 3-*D* geomagics are too intricate for comfort, for which reason I suffice with this single example.

I began above by admitting being unable to see any alternative to that of a brute-force search by computer in seeking for a 3-*D* geomagic square. Happily however, there exist less myopic individuals. Such a one is Aad van de Wetering of Drieburgen, a recognized name in recreational math circles in Holland. Figure 21.4 shows the marvellous 3-*D* square with 3×3×3 cubic target he has produced without recourse to a computer. Note that piece sizes form the consecutive series of odd numbers 1, 3, 5, . . . , while the piece shapes reflect a textbook application of Lucas's formula, shown alongside. Let his achievement be an encouragement to others.

22 Alpha-Geomagic Squares

Some readers may be familiar with the the so-called *alpha*magic square, a playful variant on the numagic theme that is really two magic squares combined in one. These are squares that use number-*words* rather than numbers, for which reason they are language-specific, English specimens being distinct from their foreign counterparts. The idea is illustrated in Figure 22.1, which is a French alphamagic square.

On the left is a square showing the value of the French number-words in numerals. On the right is one showing the number of letters they contain: 5 in *Douze*, 6 in *Quinze*, 7 in *Dix-huit*, and so on consecutively up to 13 in *Deux cent douze*. The two smaller squares are of course both magic, showing constant sums of 336 and 27, respectively. It is this property that is meant by decribing the centre square as alphamagic. Note that alphamagics need not always exhibit consecutive letter-counts, a special feature of this French example not shared by alphamagics in general.

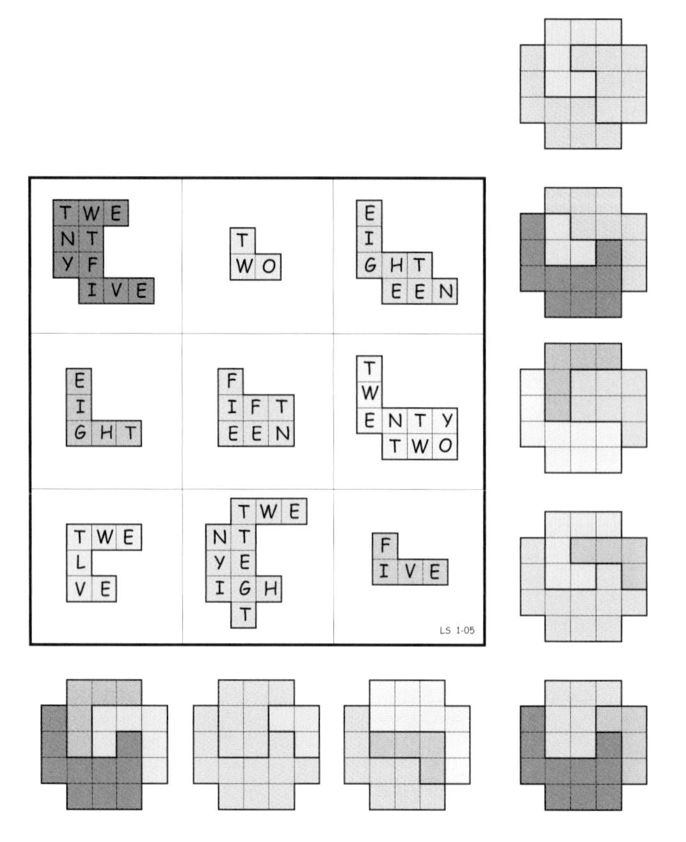

Fig 22.2 A 3×3 alpha-geomagic square.

The idea of combining an alphamagic square with a geomagic square may seem far-fetched, but can arise naturally after long gazing at targets built up from polyominoes. The desire to inscribe characters in the little squares can become overpowering. Figure 22.2 shows a 3×3 alpha-geomagic square based on a much publicized English exemplar, again showing consecutive letter-counts. Piece sizes in the geomagic square sought were thus

15	206	115
212	112	12
109	18	209

Quinze	Deux cent six	Cent quinze
Deux cent douze	Cent douze	Douze
Cent neuf	Dix-huit	Deux cent neuf

6	11	10
13	9	5
8	7	12

Fig. 22.1 A French alphamagic square.

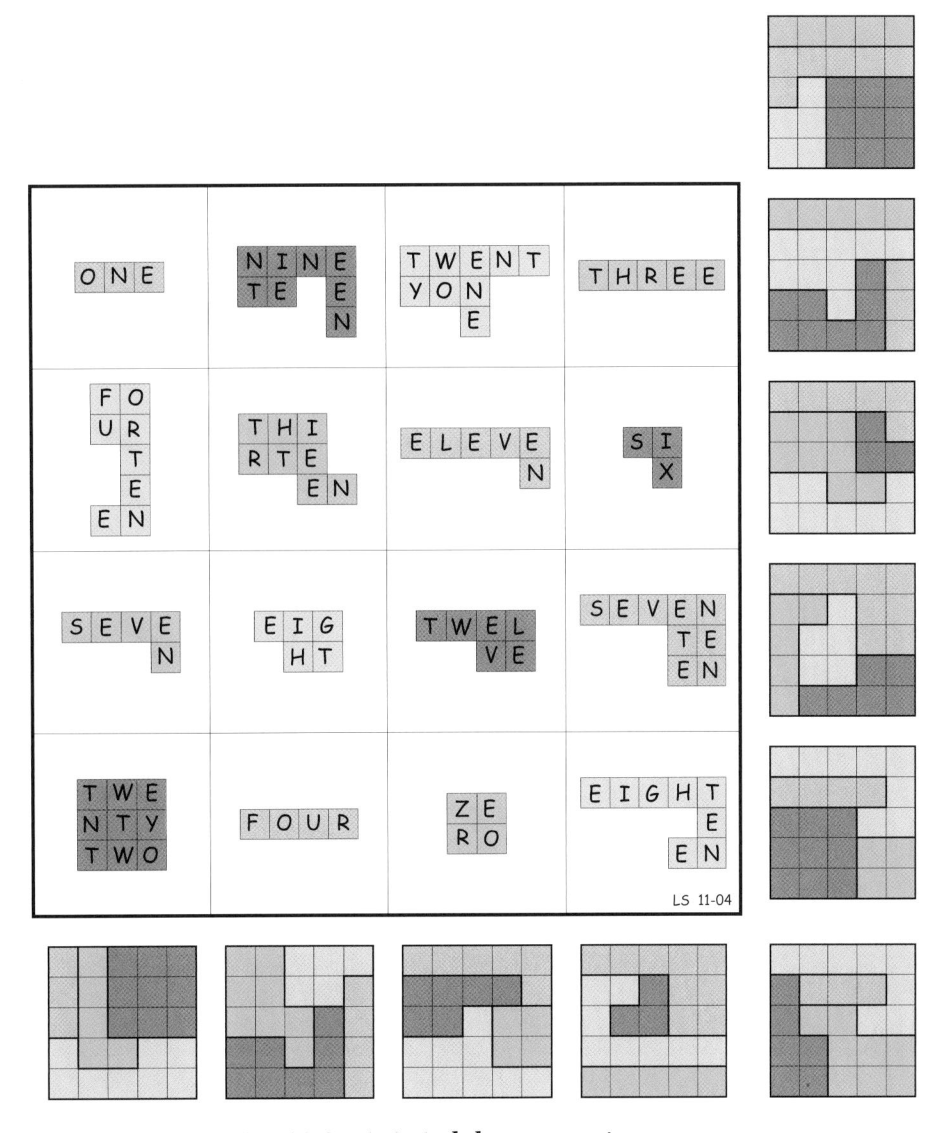

Fig. 22.3 A 4×4 alpha-geomagic square.

determined by letter-counts in the alphamagic square. The 7 letters in the centre number-word, *fifteen*, meant that the target must be of size 3×7 = 21. I was lucky enough to find a solution using this pleasing cross shape. Any three shapes in a straight line tile the cross, their associated numbers adding to 45.

Figure 22.3 shows an example of size 4×4. Once more fortune smiled on the search for a 5×5 square target to accommodate a constant sum of 25 letters contributed by the alphamagic square from which I started. Figure 22.4 shows an alternative version of the same square.

23 Normal Squares of Order 4

In Part II we looked at various aspects of 4×4 squares but without paying any special regard to their corresponding area squares. The latter are, of course, numerical magic squares in their own right, albeit ones in which the appearance of repeated values tends to discourage interest. An exceptional case that attracts special notice is thus that of the *normal* square in which the areas of the pieces form the arithmetic progression 1, 2, . . , 16.

How do we go about constructing a normal 2-*D* square? Since the area square is itself normal, it can only belong to one of Dudeney's twelve Types. This suggested the idea of starting with a numerical magic square, the formula for which is then consulted in order to arrive at the key and keyhole *areas* that will be required to construct a 2-*D* version of the same square. An example will clarify this process.

Figure 23.1 shows 'Melencolia I,' a well-known engraving by the German Renaissance artist Albrecht Dürer. An allegorical composition, it has been the subject of many interpretations. Among other symbols, it includes

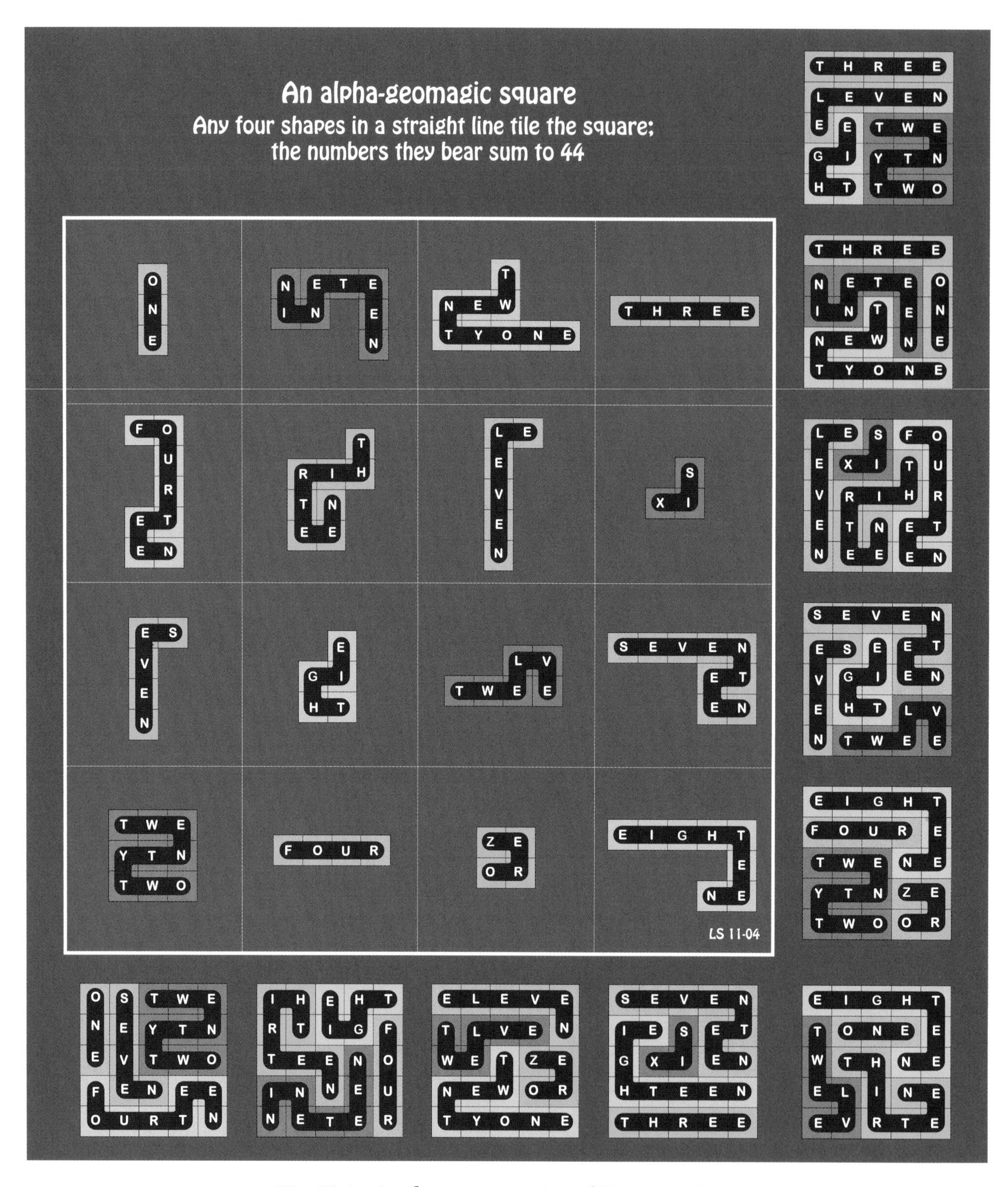

Fig. 22.4 An alternative version of Figure 22.3.

Fig. 23.1 'Melencolia' by Albrecht Dürer.

16	3	2	13
5	10	11	8
9	6	7	12
4	15	14	1

Fig. 23.2 A clearer view of Dürer's square.

a 4×4 magic square using the numbers 1 to 16, with constant sum 34. A point of interest is that the date of the engraving, 1514, is given by the two centre numbers in the bottom row. Here then is a normal numerical magic square that will serve to illustrate the creation of a normal 2-*D* square. Figure 23.2 shows a modern version of Durer's square that is easier to read.

A look at the distribution of the complementary pairs in Figure 23.2 shows them balanced about the centre in a pattern corresponding to the Dudeney Type III diagram. Figure 23.3 reproduces the generalization of Type III squares, seen earlier in Figure 13.2 on page 38. The values taken by the variables in the case of Durer's numerical square are then easily derived:

From $(A + p + q) - (A - p + q) = 2p$, we find $16 - 10 = 6$, or $p = 3$,
From $(A + p + q) - (A + p - q) = 2q$, we find $16 - 7 = 9$, or $q = 4\frac{1}{2}$,

$A+p+q$	$A-q-r$	$A-p-s$	$A+r+s$
$A-q+r$	$A-p+q$	$A-r+s$	$A+p-s$
$A-p+s$	$A+r-s$	$A+p-q$	$A+q-r$
$A-r-s$	$A+p+s$	$A+q+r$	$A-p-q$

Fig. 23.3 A generalization of Dudeney Type III squares.

From $(A + r + s) - (A - r + s) = 2r$, we find $13 - 11 = 2$, or $r = 1$,
From $(A + r + s) - (A + r - s) = 2s$, we find $13 - 6 = 7$, or $s = 3\frac{1}{2}$.
Similarly, from the magic sum $4A = 34$, we find $A = 34 \div 4 = 8\frac{1}{2}$.

In constructing any 2-*D* square, the first thing to be decided upon is the target shape. Now the target area in this case is 34, which is not a square number. A straightfoward square target is thus ruled out. My own liking being strongly for square targets, the idea of excising two cells from a 6×6 square outline was then an obvious next best choice: 6×6 – 2 = 34. The latter target with its two holes can be seen in Figure 23.4, which is central to the discussion to follow.

Returning now to the template in Figure 23.3, we see that the array to be detrivialised is a uniform square of *A*'s, with magic sum 4*A*. This implies an initial 2-*D* substrate showing 16 identical pieces, four of which must assemble to form the target. Can the chosen target be dissected into 4 congruent pieces showing areas of 8½, as required? The 4 red 'notched' triangles labelled '*A*,' drawn as if on squared paper, at the top of Figure 23.4, show that it can. Accordingly, we begin with a 4×4 array consisting of 16 of these notched triangles, one in each cell. In Figure 23.4, the latter are drawn in red, their notches coinciding with a small darker square showing the position of the neighbouring hole within the target. Shapes outlined in black indicate the finished pieces resulting from additions to and/or excisions from the initial notched triangles. These same piece shapes, all of them polyominoes, reappear in the targets (slightly reduced in size) to right and below.

This brings us to the ticklish matter of determining the key and keyhole *shapes* to be used. Take for example the top left hand entry, corresponding to $A + p + q$ in the template, or the number 16 in Dürer's square. The areas of *A*, *p* and *q*, have already been established. They are 8½, 3 and 4½, respectively. The shape of *A* is the notched triangle, but what shapes should *p* and *q* take? Assuming rectilinear shapes, and given the constraint on their areas, there are not very many possibilities. In appending them to the notched triangle, they must form a piece with an area of 16. But *p* and *q* also play a part in the formation

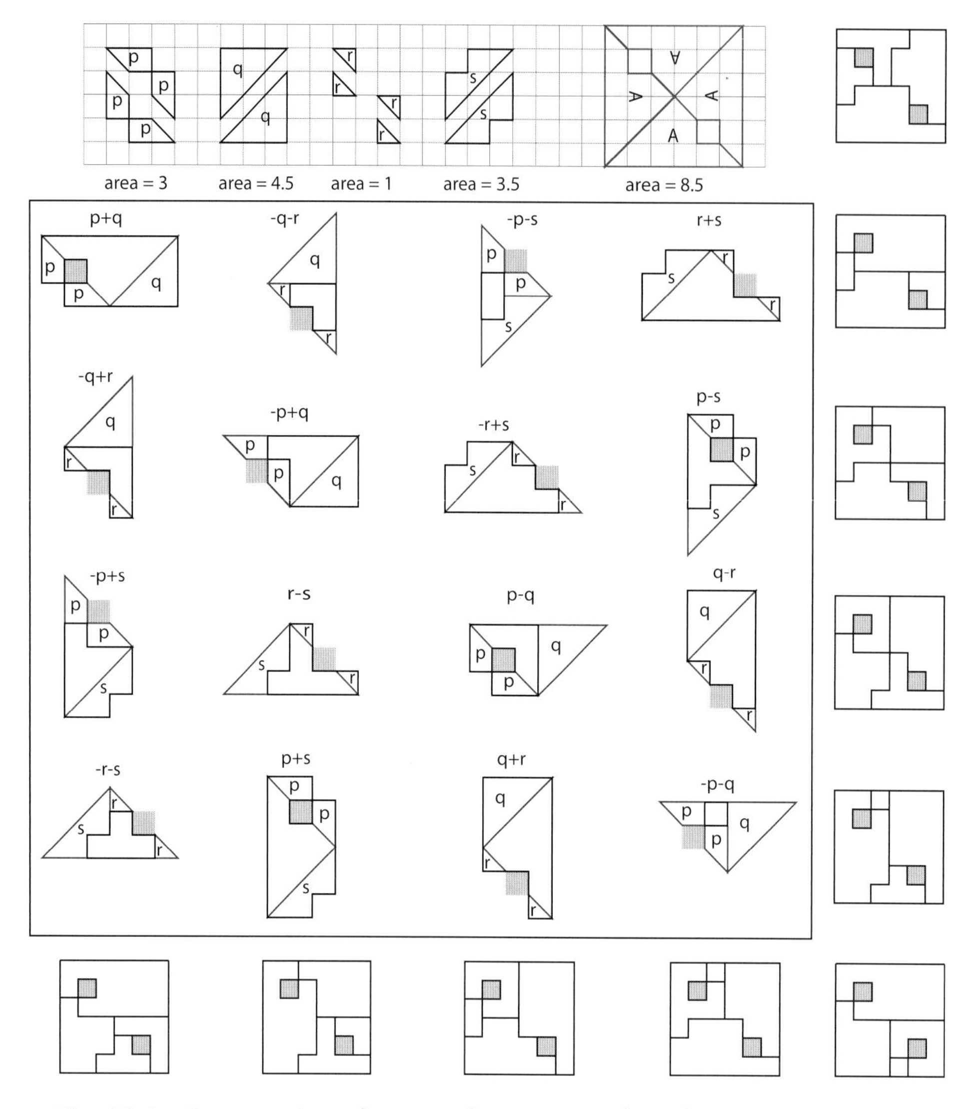

Fig. 23.4 Construction of a normal 2-*D* square based on Durer's square.

of several other pieces in the square under construction. For example, consider the piece corresponding to $A - p - q$ in the bottom right-hand cell. Here, the same p and q shapes that augmented A in the top left-hand cell must now be excised from A to leave an area of unit value: the number 1 in Dürer's square. Continuing in the same vein, different candidate shapes for p and q can be tried out in these and still other pieces until success in all cases is achieved. Or not, of course, depending upon whether the target chosen does or does not admit of solutions. In any case, it is a matter of patient trial and error. Similar remarks apply to the selection of shapes for the variables r and s.

An important personal insight that is incorporated in the example of Figure 23.4 lies in its use of both *weakly-connected* and *disconnected* key shapes. For a long time, I suffered from an unconcious assumption that keys and keyholes must necessarily be formed of *connected* areas. The effect of this blind spot was to thwart every attempt to produce a normal square, so that the awakening from this illusion, when it came, was a sudden release into sunshine followed by a spate of new discoveries. The latter can be seen in Figures 23.5 – 23.11, all of which are normal squares of a distinct Dudeney Type. The increasingly flamboyant target shapes that emerge are a reflection of the author's improving grasp of the sometimes intricate techniques involved.

There may be readers who would like to create normal 2-*D* squares of their own. If so, I can only encourage them to avoid the kind of pitfall just described, through a careful examination of Figure 23.4, in which I have done my best to elucidate the method of construction. At top, above the main geomagic square, are seen the 4 shapes used for the keys corresponding to variables p, q, r, and s. The square grid on which they are drawn makes it easy to verify their respective areas. Squared paper,

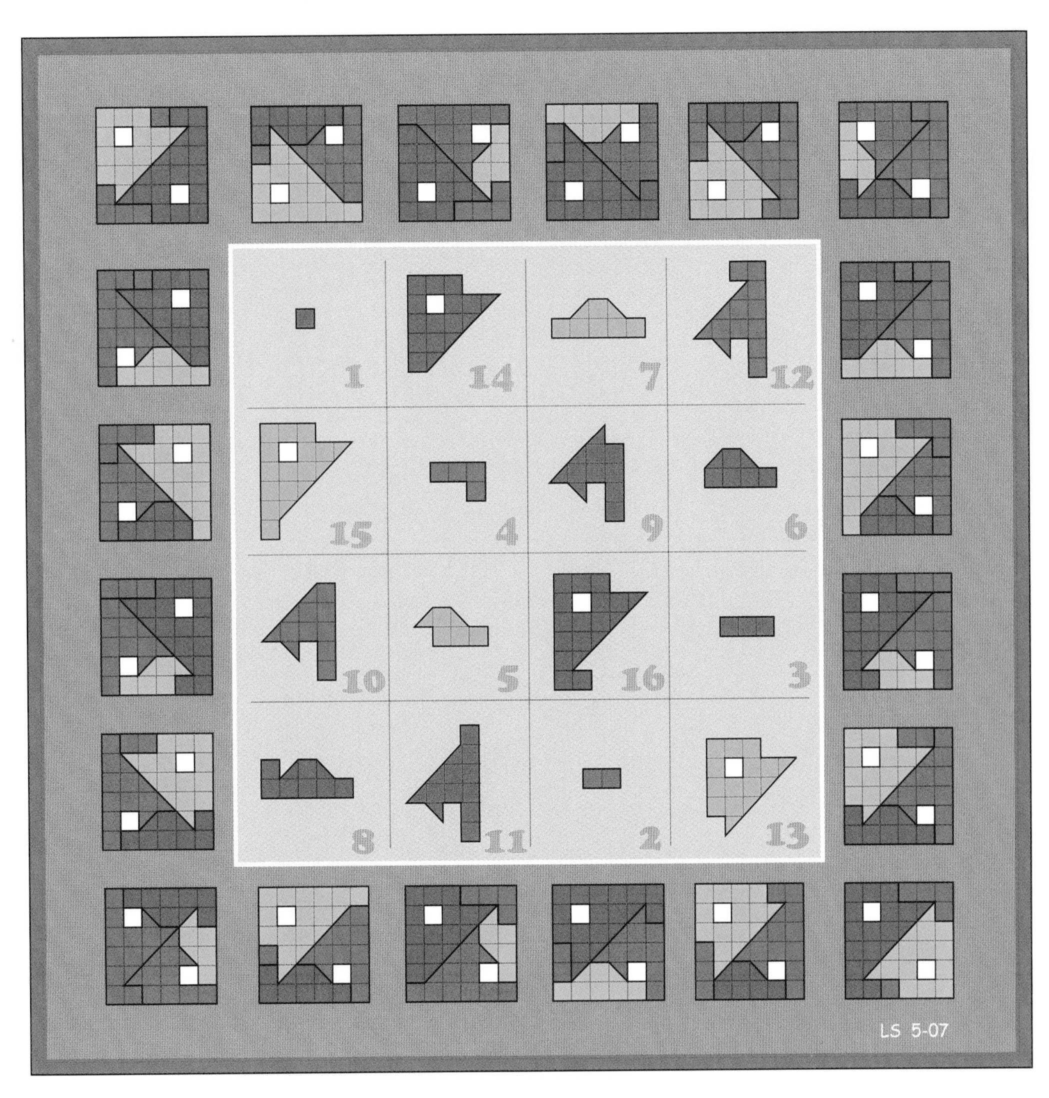

Fig 23.5 A Type I square again using a 6×6 square target with two symmetrically disposed cut-outs. As with the squares in the Figures to follow, with a little patience, dedicated readers ought to be able to deduce the initial substrate shape used, along with the keys and keyholes that give rise to the individual piece shapes.

I might add, along with a pencil and eraser (as well of course, as a magic wand) are the indispensible tools of the practicing geomagician. For ease of comparison, each key is shown in those two orientations in which it appears within the geomagic square beneath. Careful scrutiny of the individual pieces in the latter will then reveal how the final piece shapes are derived from the notched triangle modified by the addition and/or subtraction of the p, q, r, s shapes. The algebraic terms above each piece indicate the operation to be performed on the triangle, as dictated by the formula in Figure 23.3. For example, in the third row from top, left-hand column, we find $-p + s$, which means, excise p and append s, to result in the nonomino or piece built from 9 squares shown. Note that the orientation of each piece has been chosen with a view to clarity, but is not otherwise of special significance.

I can hardly conclude this discussion of normal order-4 squares without mentioning a discovery of some interest.

As already mentioned, among the more familiar landmarks dotting the long history of magic squares is Frénicle's exhaustive enumeration of the 880 normal numagic squares of order-4, first published in 1693. This total of 880 excludes those seven trivial variants of each square resulting from rotations and/or reflections. Frénicle's result has often been verified, so that the figure of 880 squares using the natural numbers from 1 to 16 is a well-known result in the literature of the subject. At the same time, there seems to have grown up a widespread belief or assumption – not always conscious perhaps – that 880 is the therefore the maximum number of different magic squares that could be formed by *any* set of 16 distinct numbers. Given the regularity of 1, 2, . . . , 16,

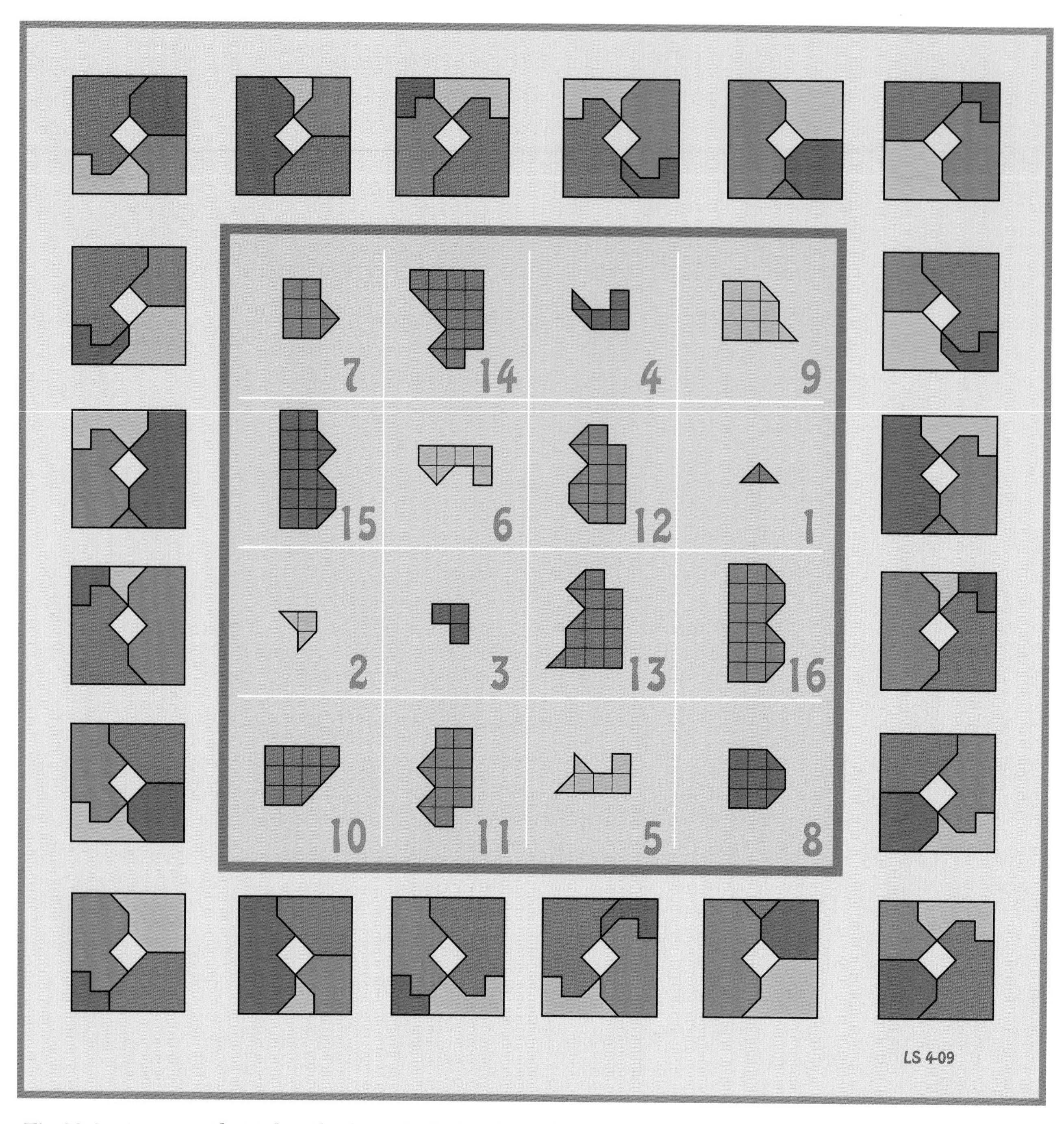

Fig 23.6 A square that takes the formula for Dudeney Type X squares in Figure 13.2 as its starting point. Recall that the target area in a normal square must be 34, a restriction that leaves the designer with little freedom of choice. I was therefore especially pleased to arrive at the solution seen here, using a square target with a rotated square hole at its centre.

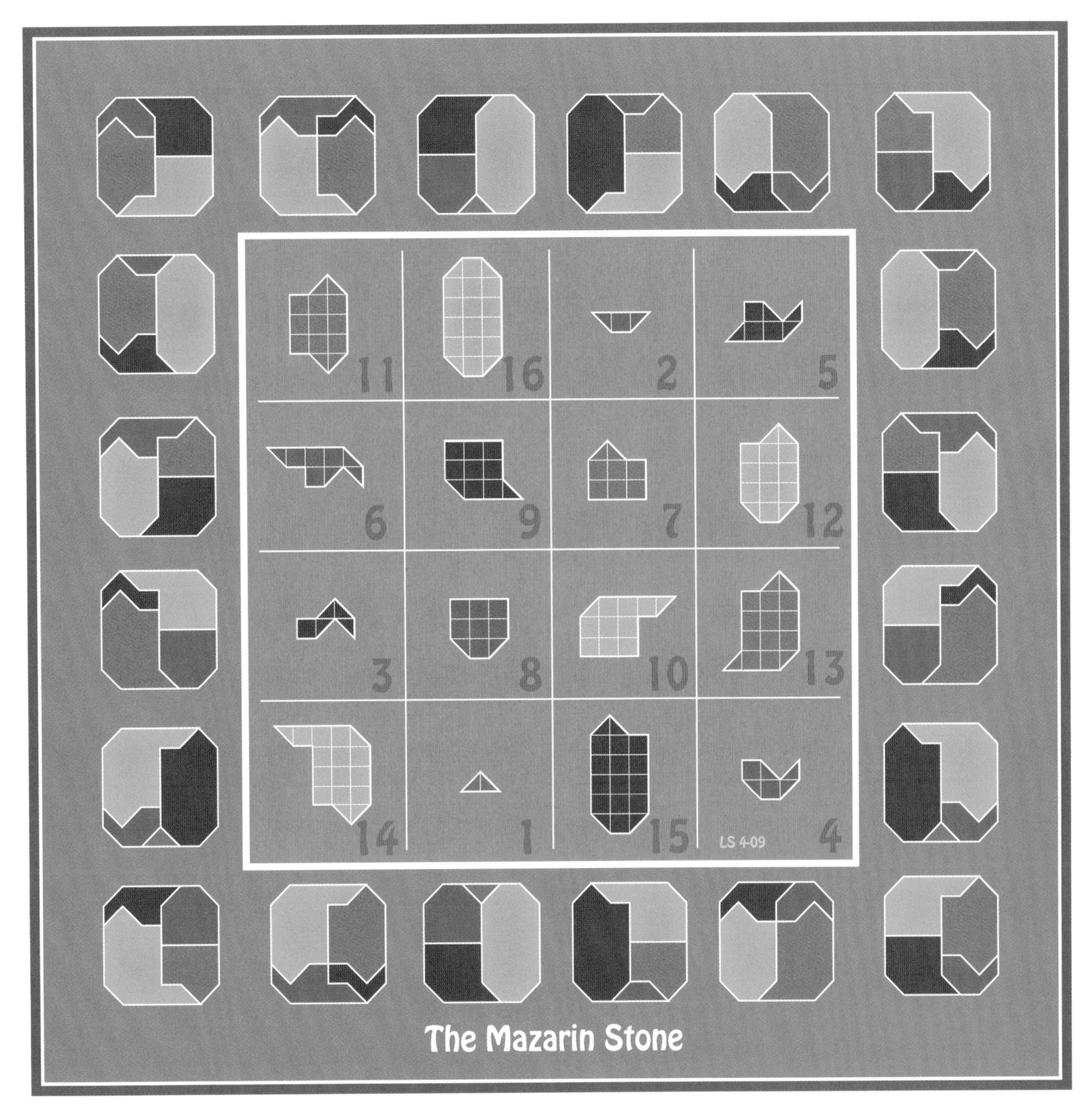

Fig 23.7 Experimenting with different Types, I took Dudeney's formula for Type IX as the template for this square. 'The Mazarin Stone' is the name of a famous yellow diamond that becomes the subject of a Sherlock Holmes investigation in an Arthur Conan Doyle story of the same name. The gem-like target shape of area 34 that has been arrived at is celebrated in the picture title.

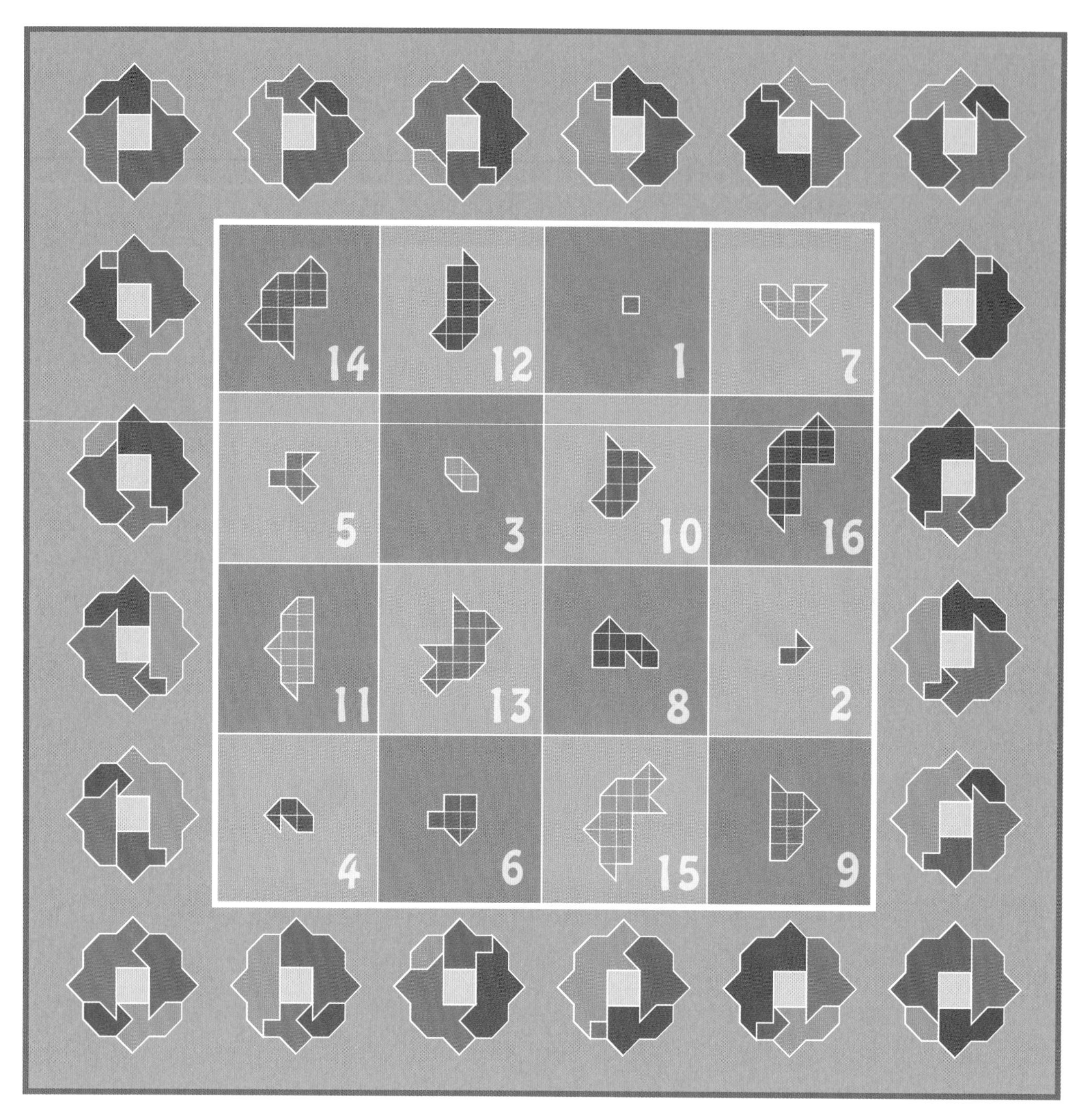

Fig 23.8 Important as is the choice of template used, one can hardly get far in designing a geomagic square before deciding on the target shape. Here, the template used is the formula for Type II in Figure 13.2, which implies a substrate of 16 identical pieces. This in turn calls for a target that can be dissected into 4 congruent pieces, a goal easiest to achieve when the shape chosen has four-fold rotational symmetry.

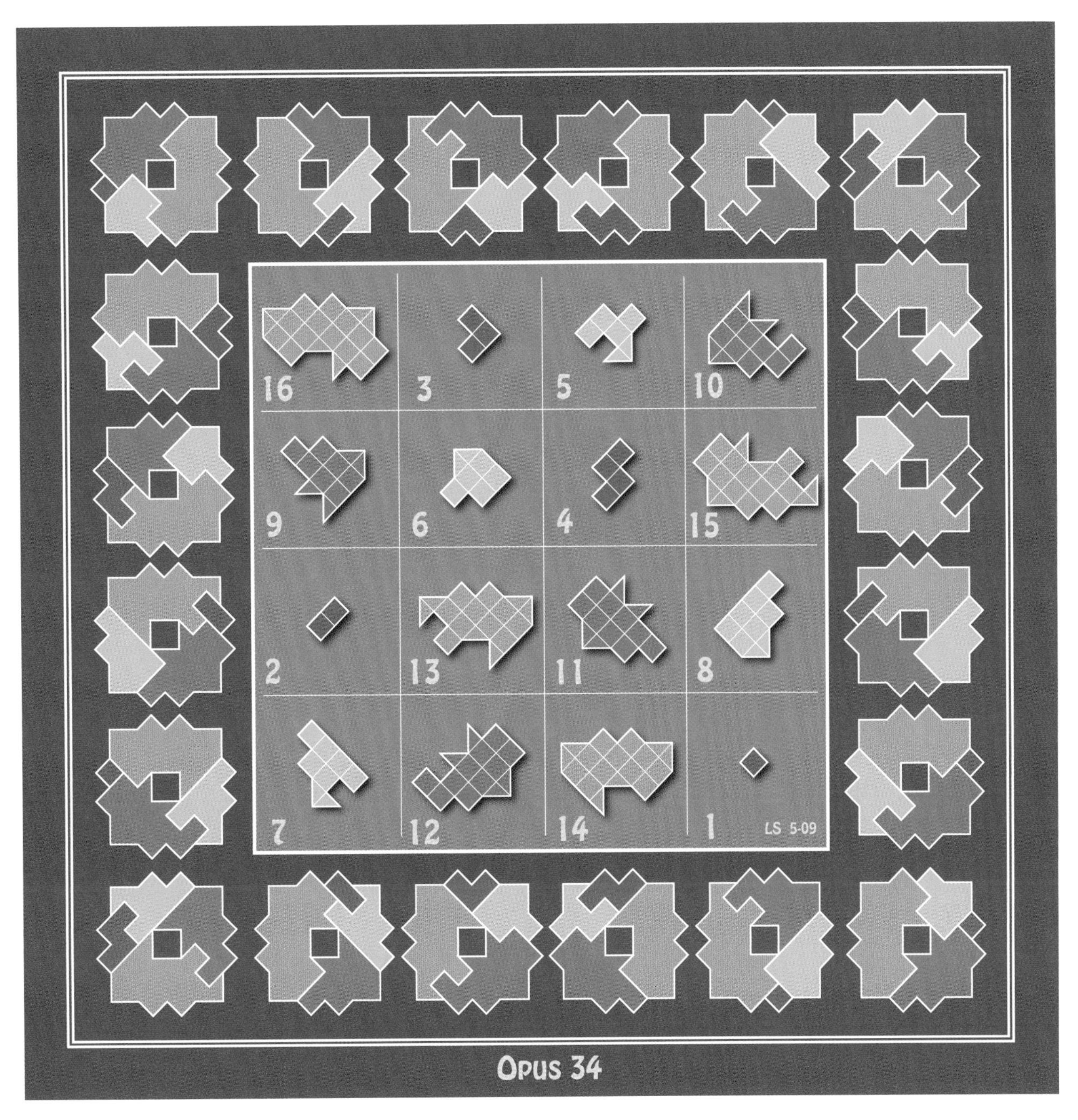

Fig 23.9 Four-fold rotational symmetry again underlies the choice of target shape in this square that is based on the Type III formula in Figure 13.2. As in the analysis of Figure 23.4, the substrate square consists of 16 identical pieces each with a area of 34 ÷ 4 = 8½. Having decided not to clutter the cells with numerals, the title 'Opus 34' is a sly pointer to the target's area, and with it to the fact that piece sizes again form the progresssion 1, 2, 3, ...

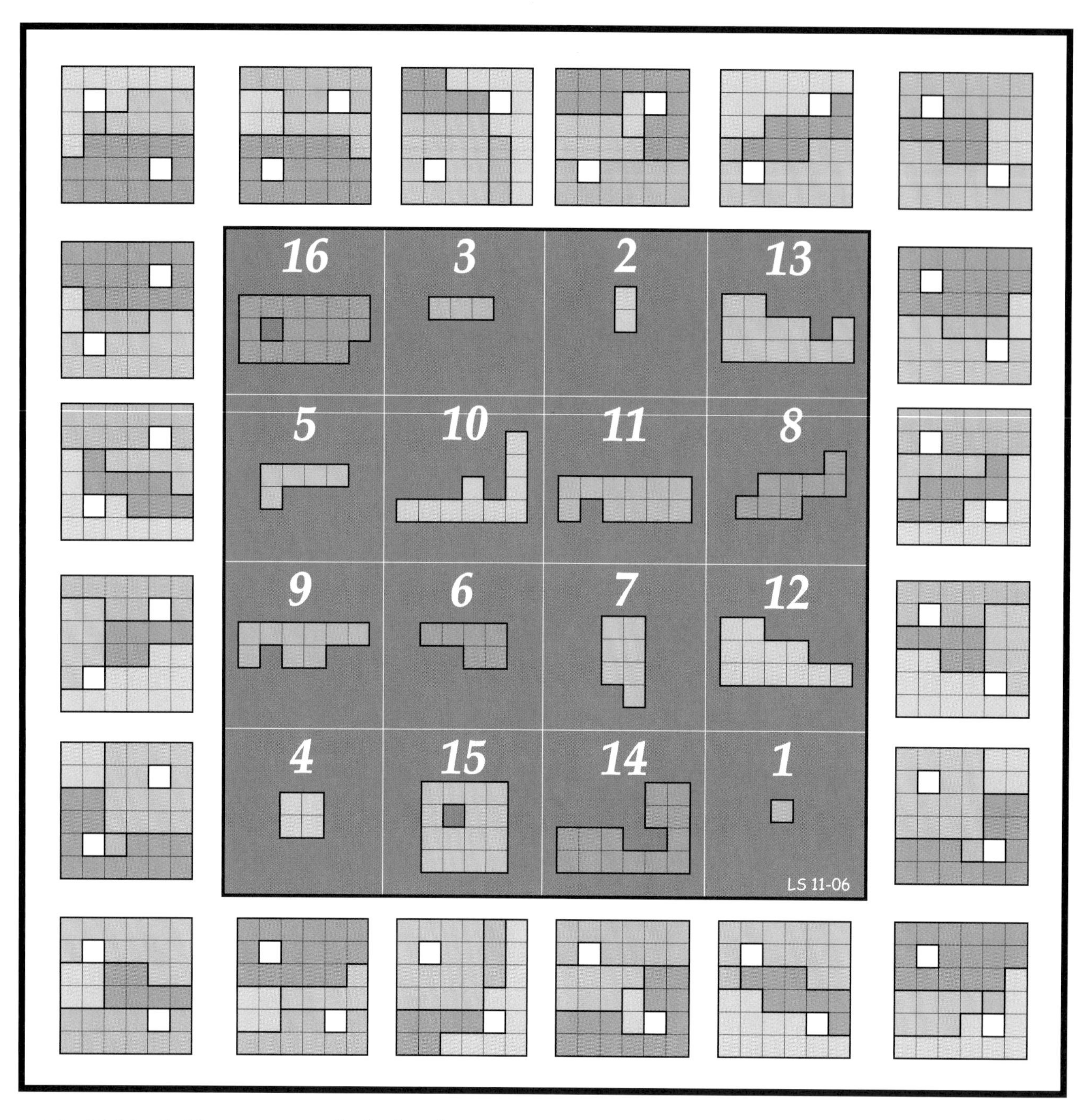

Fig 23.10 In Figure 23.4 we looked at the construction of a 2-*D* square based upon Dürer's numerical square. Here we see an alternative geometrical version discovered by computer. An ambitious project I am saving for my old age is to produce a further 2-*D* version of Dürer's square in which the entries for 15 and 14 would be (disconnected) pieces bearing a close resemblance to these very same numerals. 1514 is, of course, the date of Dürer's original engraving.

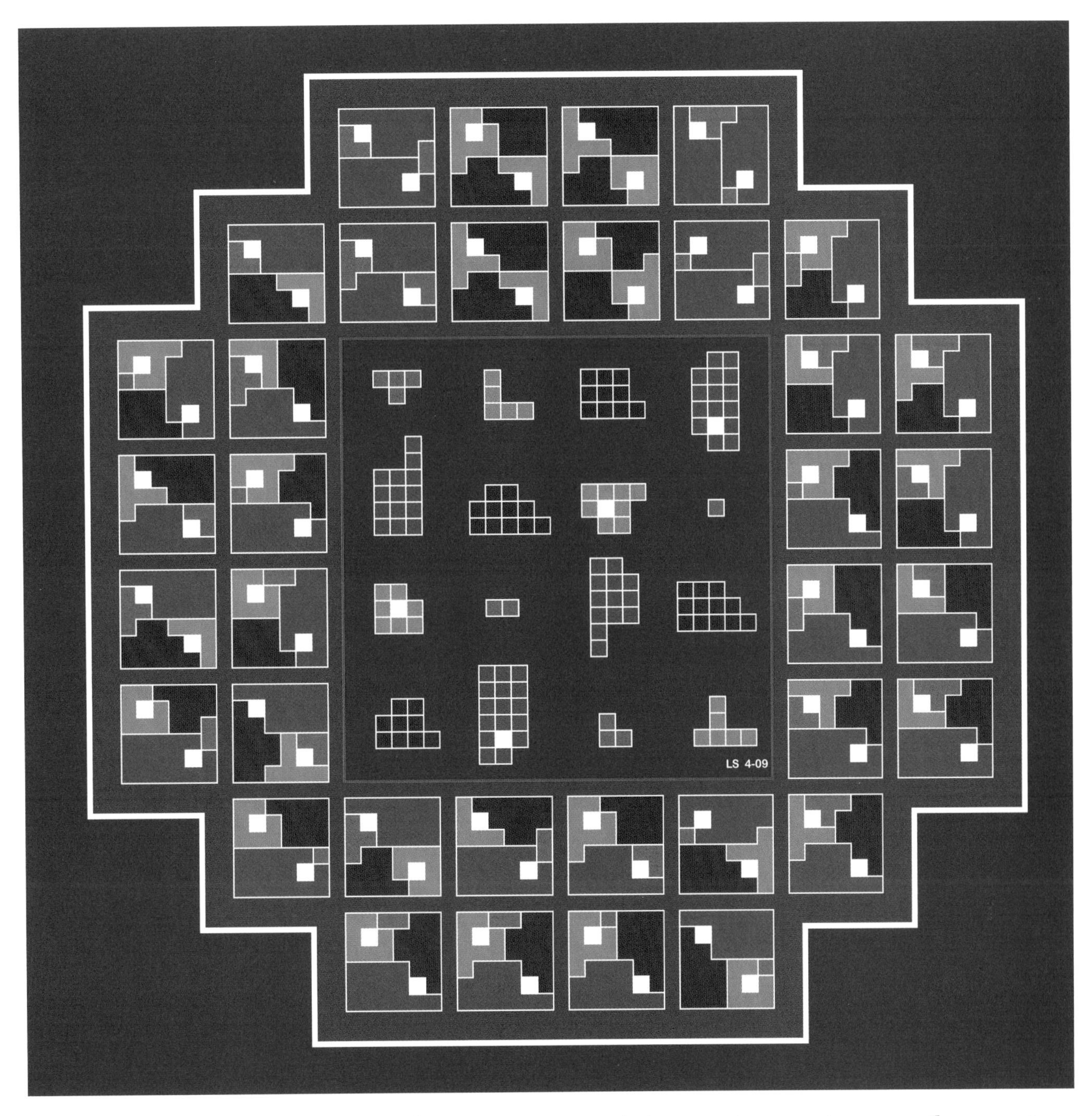

Fig 23.11 In the realm of numerical magic squares, any 4×4 panmagic square is necessarily compact and any compact square necessarily panmagic. But not so the world of 2-*D* magic squares. The 4×4 square here is compact but it is not panmagic. Similarly, there exist 2-*D* squares that are panmagic without being compact; see, for example, Figure 16.8.

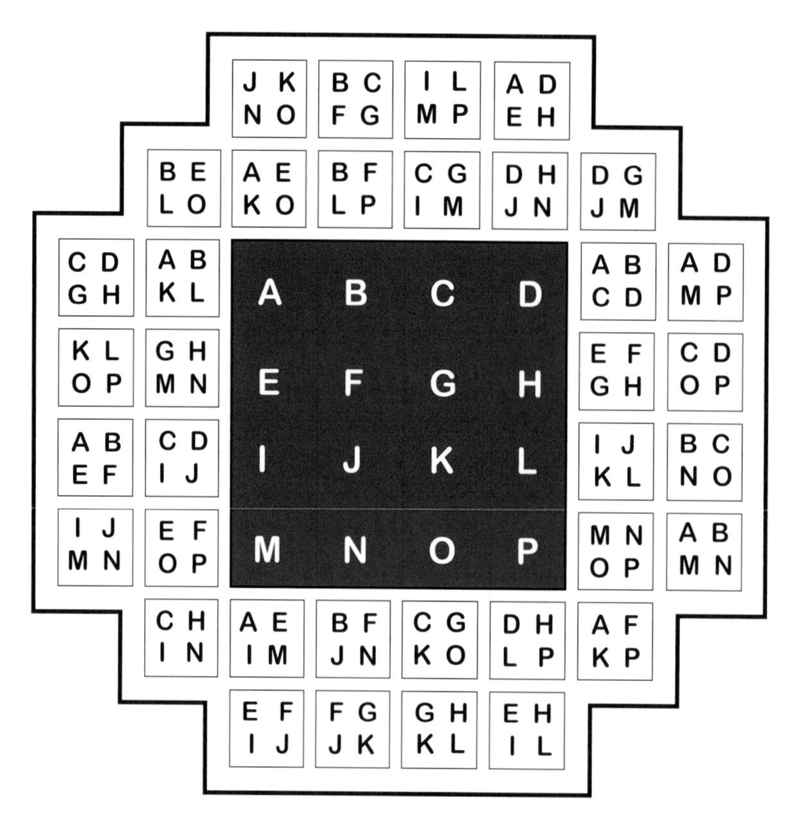

Fig 23.12 A key to Figure 23.11.

12	5	11	8
2	17	3	14
7	10	6	13
15	4	16	1

Fig 23.13 One of 1040 numagic squares using 1 – 17, but no 9.

1	2	3	4	5	6	7	8	9	10	11	12	Type
48	48	48	96	96	480	52	52	52	52	8	8	Total

Table 5 A breakdown of the 1040 squares into Dudeney's 12 Types.

forming as they do a simple arithmetic progression, such an assumption will have seemed natural enough to many.

However, in 1996, I wrote a computer program able to generate every possible 4×4 magic square constructable from any given set of 16 distinct integers. An early outcome of this enterprise showed the above assumption to be untrue. Extensive trials have revealed the existence of one particular set that yields 1040 rather than 880 squares. It is a balanced or 'palindromic' set of integers that do not form an arithmetic progression: – 8, – 7, . . . , –1, 1, 2, . . . 8. Note the absence of zero. The value of the new magic sum is also zero. Being built up from 8 complementary pairs (each of which again sum to zero), these 1,040 squares can again be classified in terms of Dudeney Types, the number of squares of each Type being as seen in Table 5.

Adding a constant to each number so as to yield a similar set, but now using 16 positive integers, will have no effect on the number of magic squares produced. Figure 23.13 shows such a square in which the smallest number is 1. Its magic sum is 36.

The previous finding is not without ramifications in the realm of 2-*D* magic squares. Recall that the formulae we have been using as templates are derived from Dudeney's twelve Types, which are themselves based on the distribution of complementary pairs of numbers. Now complementary pairs have already been encountered in the section on 3×3 squares, where several examples were shown of piece pairs that are the *geometric* complement of each other. By which I mean pairs that will fit together so as to complete an identical shape in each case. Figure 6.7 on page 14, for

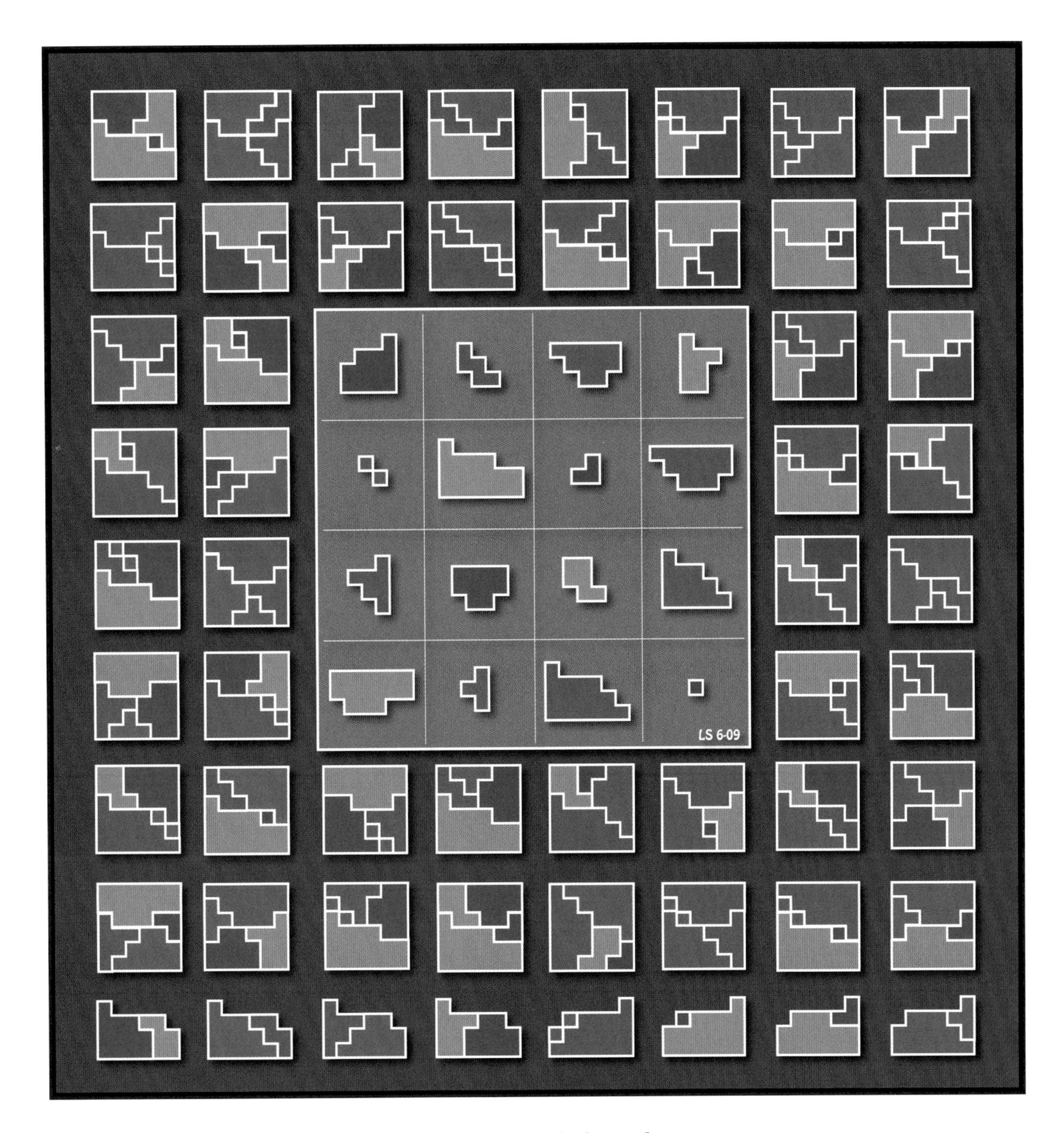

Fig. 23.14 A 4×4 square composed of complementary piece pairs.

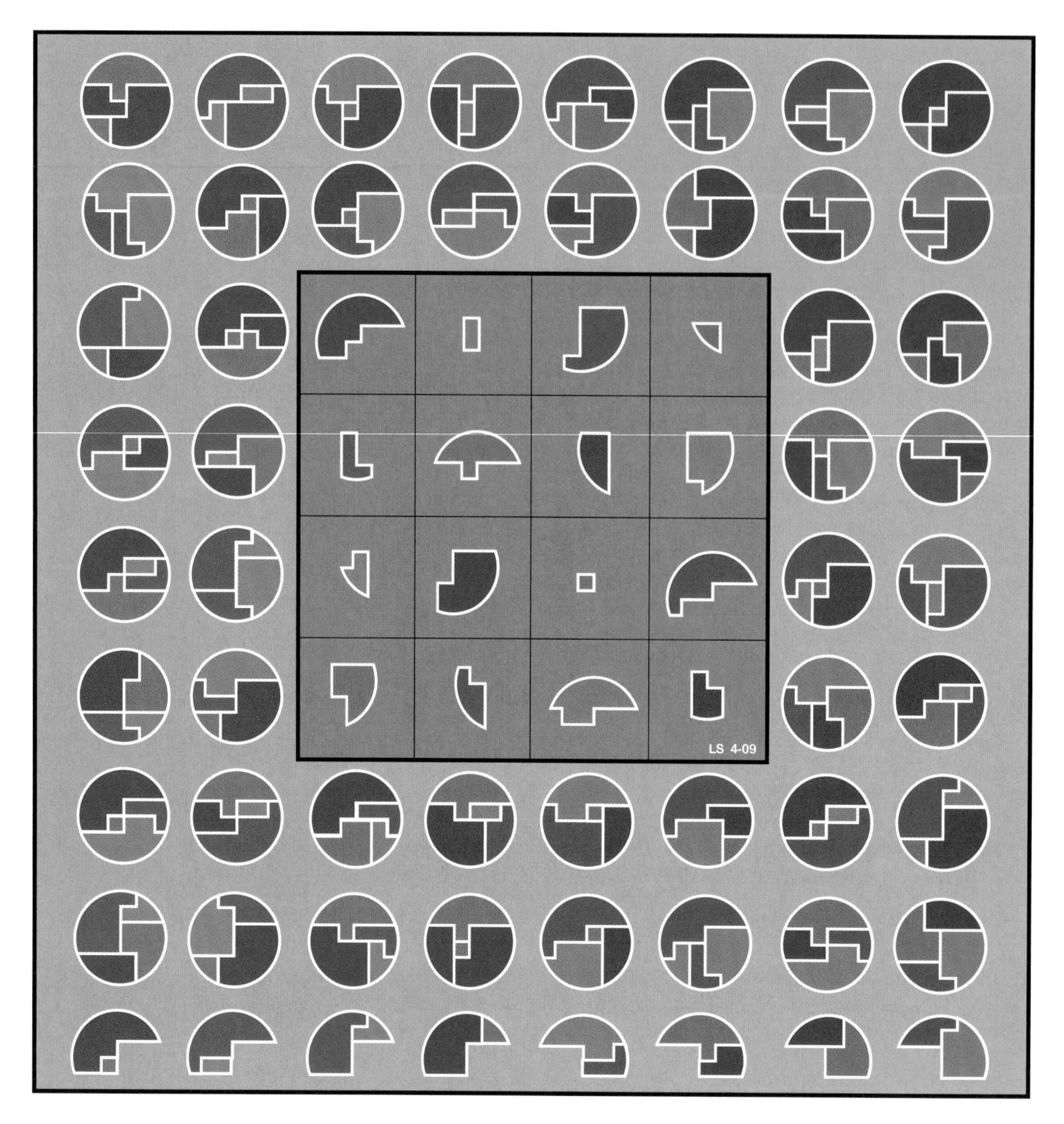

Fig. 23.15 A square similar to Figure 23.14 but now with circular targets.

example, was produced with the purpose of illustrating this property. Thus far, however, we have not met with a single instance of a 4×4 square composed of piece pairs that are able to perform the same trick.

I have to admit that, for reasons as yet obscure, attempts to produce a *normal* square with this property have not met with success. However, in Figures 23.14 and 23.15 can be seen two such examples based not upon normal numagics but on Fig 23.13, which is one among the 1040 squares that can be formed using the numbers 1 to 17, excepting 9. The two figures are exact analogues, their only differences lying in the choice of piece and target shapes. In both, the 16 pieces are composed of 8 complementary pairs. Every pair tiles an identical shape that is one-half of the target. The pairs can be seen in the bottom row of figures, any two of which will combine to complete the target. In all, there are 52 groups of four target-tiling pieces, only 48 of which are shown. The remaining four can be assembled using the four corner pieces belonging to each of the four embedded 3×3 sub-squares. It can be shown that these same sets of 16 pieces admit of rearrangements yielding a total of 528 distinct

2-*D* squares, rotations and reflections not counted. A key to both Figures can be seen in Figure 23.16.

Four distinct 4×4 geomagic squares occupy the corners of 'Art of Fugue' in Figure 23.17: one yellow, one blue, one orange, and one brown. All four squares are both nasik and compact. The set of sixteen pieces used is the same in each case. Although not immediately apparent, they comprise eight pairs, each of which will again combine to tile an identical shape that is half the target. At the centre of the picture, slightly enlarged copies of the four corner squares (minus their targets) are shown superimposed, their pieces now revealed as exactly coinciding with each other so as to pave the 16 four-colored cross shapes. For example, the top left-hand cross in the central 4×4 array is formed by the four pieces belonging to the top left-hand cell of each corner square. Similarly shaped targets surround this ensemble. Figure 23.18 provides a key to the composition of targets.

DF JP	BC IL	JK NO	BC FG	IL MP	AD EH	BH LN	AD JK
EH NO	AG KM	AH KN	BE LO	CF IP	DG JM	FG MP	CE IO
CD GH	AF KP	A	B	C	D	AB CD	AD MP
KL OP	BG LM	E	F	G	H	EF GH	CD OP
AB EF	CH IN	I	J	K	L	IJ KL	BC NO
IJ MN	DE JO	M	N	O	P	MN OP	AB MN
AE KO	BF LP	AE IM	BF JN	CG KO	DH LP	AB KL	CD IJ
CG IM	DH JN	EF IJ	FG JK	GH KL	EH IL	EF OP	GH MN

Fig. 23.16 Key to Figures 23.14 and 23.15.

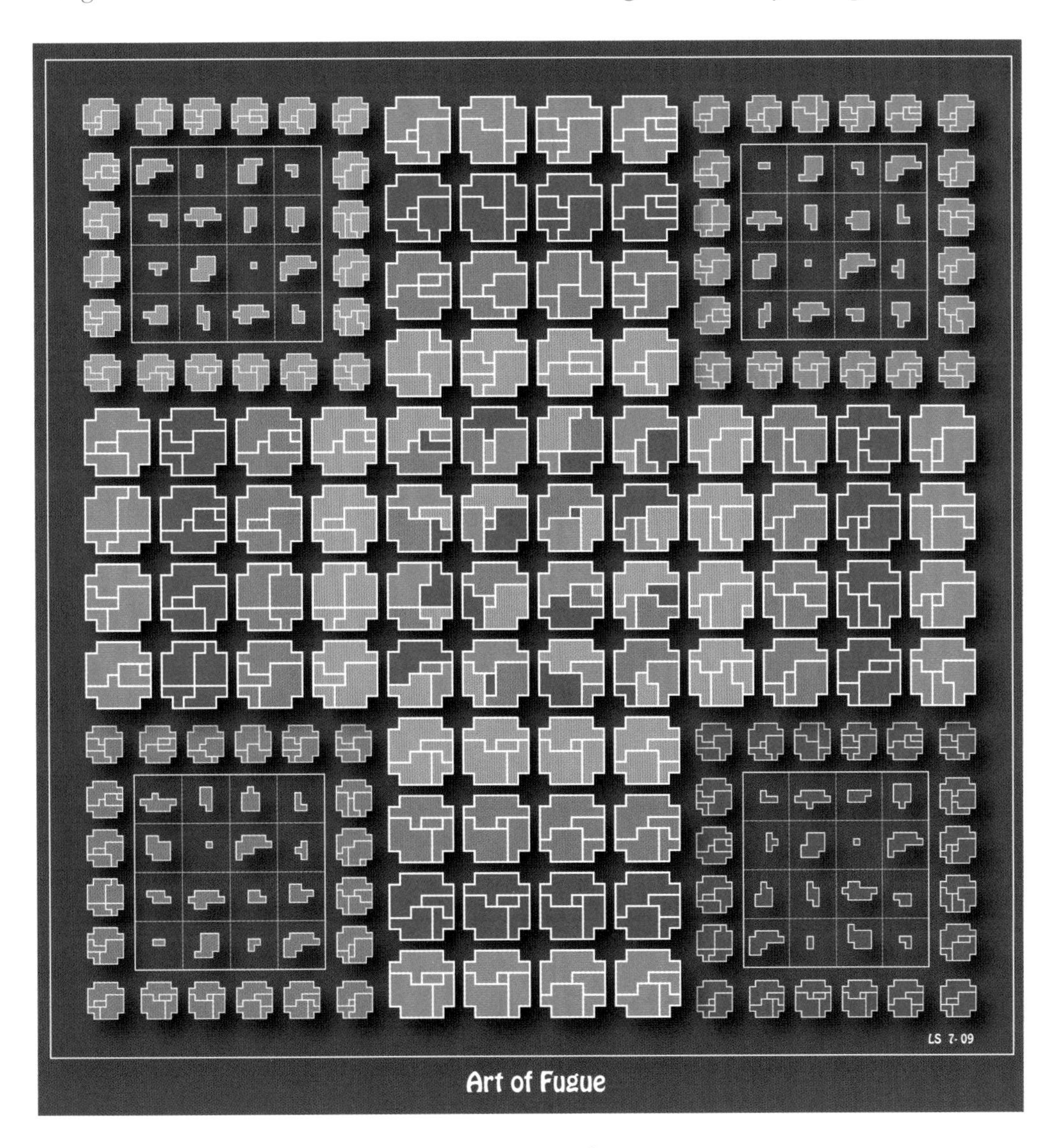

Fig. 23.17 'Art of Fugue'.

24 Eccentric Squares

If there is one thing that has emerged during our journey through the geomagic jungle it is that the local fauna is far more variegated than that met with in the realm of numerical magic squares. It is not just that the lions and tigers hereabouts are more colorful and roar louder; there exist creatures in Flatland that have no counterpart in a one-dimensional world. As a result, 2-*D* magic squares are not always easy to classify in terms of established categories. This can create problems for the author bent on presenting a coherent account, whose only remaining course is to set out his stall and hope for the best. To this end, in the following I present a selection of special finds that are obviously worthy of attention, if difficult to place within in a logical scheme.

A good title for Figure 24.1 might be 'Binary Star', because the starry blue structure at the centre of the square can be interpreted as a 3×3 array in two different ways:

A B C	***B C F***
D E F	***A E I***
G H I	***D G H***

As the targets show, *both* are 3×3 geomagic squares. Note that this ambitious goal has been achieved at a price. The pieces used are all pentominoes, some connected, but mostly weakly-connected or completely disconnected. Naturally, I would have preferred them all to be connected, but no such solution could be found. Likewise, the target shape is not all that desire would wish. Nevertheless, the square is a nice instance of the

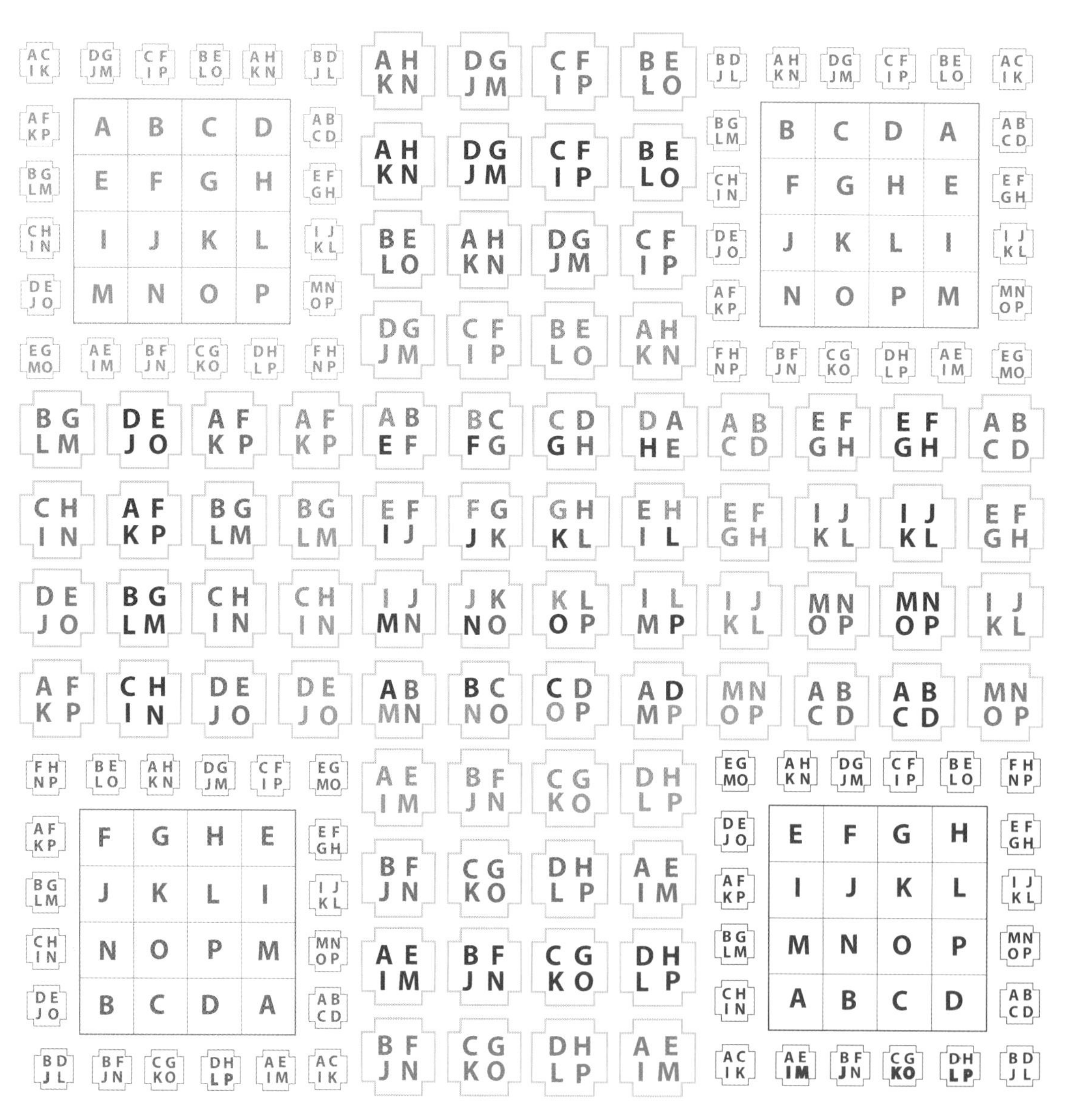

Fig. 23.18 Key to 'Art of Fugue'.

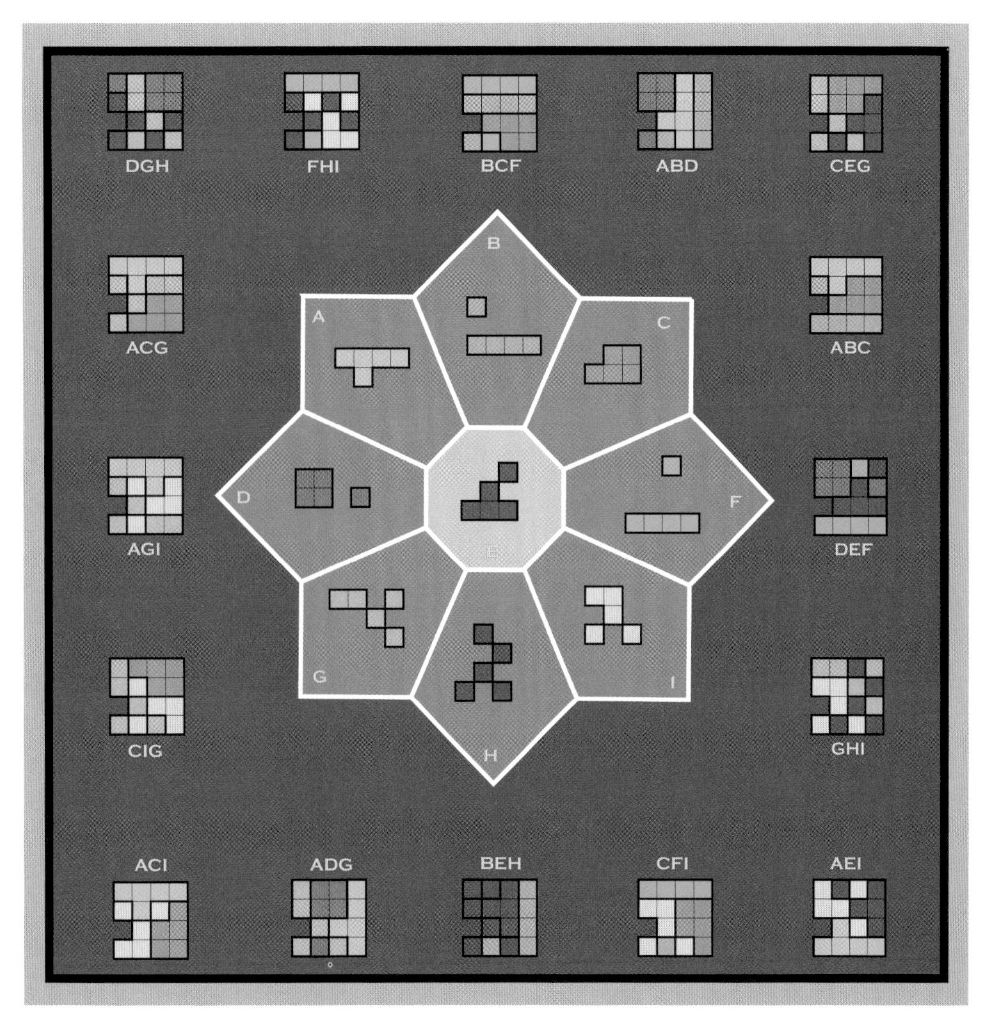

Fig. 24.1 'Binary Star:' two geomagic squares in one.

more potent magic structures that can be found in the world of two dimensions.

Previously I mentioned lions and tigers, but these are not the only animals roaming the forest. Figure 24.2, 'Genetic Engineering,' shows a swarm of butterflies collected amid the same woodland glades. Together with the picture to follow, this square celebrates a breakthrough. In creating a geomagic square, the choice of admissible target shapes is usually tightly constrained. Scrutiny will show that the target here is essentially a simple chain of four pieces. But two of the pieces, both yellow, always occur at the ends of the chain, which means they can be extended so as to take on almost any shape, without introducing deleterious side effects. Here they have been extended so as to meet, and so completely envelope the chain itself, the shape of their combined outline then being entirely up to the constructor. The choice of a butterfly struck me as an improvement over that of the triangle used in Figure 11.8 ('Indian Reservation') on page 38, which embodies the same principle of construction.

Fig. 24.2 'Genetic Engineering.' The target is essentially a chain of 4 pieces.

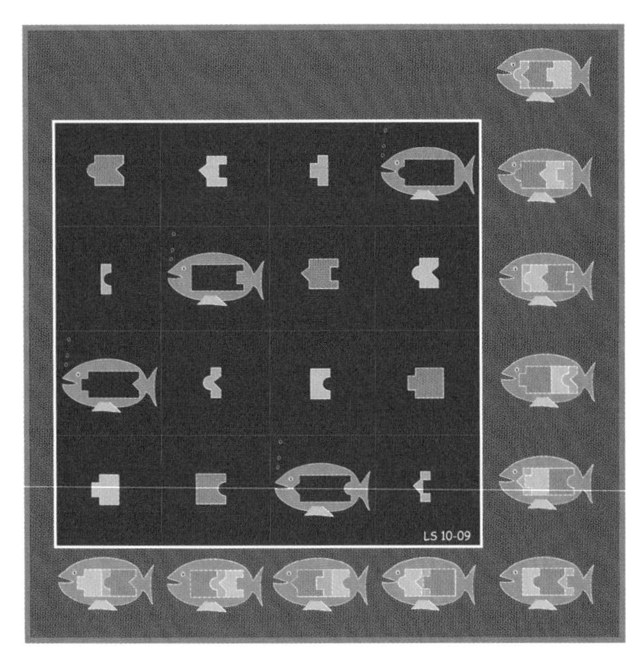

Fig. 24.3 The target is a again a chain, but now *closed.*

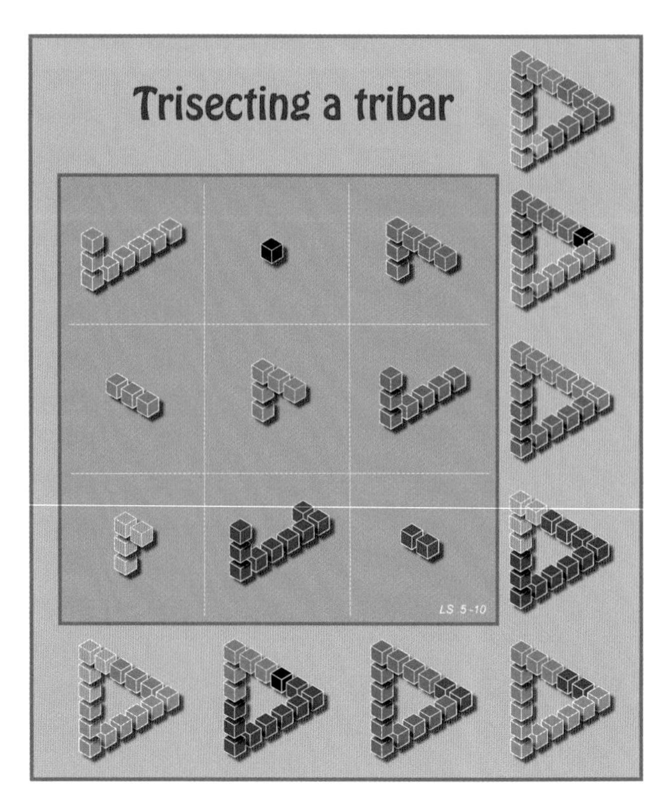

Fig. 24.4 An interesting case. But is it a geomagic square?

The envelopment principle used to create butterflies soon gave rise to a related notion, which was that of a chain of pieces nested within a *single* piece, rather than enveloped by two. That the choice of target shape remains arbitrary then becomes still more obvious to the viewer, as the school of fishes in Figure 24.3 attests.

In contemplating a new geomagic square, it was often the idea of achieving some particular target that formed the starting point of investigations. Figure 24.4 shows the most memorable instance of this process.

The target here is a variation on the first ever *impossible figure* that depicted a similar object, but one using nine rather than fifteen cubes. The latter was invented by the Swedish artist Oscar Reutersvärd in 1934. In 1980 three of his designs were depicted on Swedish postage stamps, a signal honour to their discoverer. Later, in 1958, Lionel Penrose and his son, the now famous Sir Roger Penrose, unaware of Reutersvärd's work, published an equivalent figure composed of three solid beams, nowadays known as the Penrose tribar. It appeared in the *British Journal of Psychology* [14]. A reflected version of the original tribar is seen in Figure 24.6. The result pictured in Figure 24.4 is one of two solutions using the same target and similar pieces composed of consecutive numbers of cubes. The idea of such a target had ocurred to me long before, but lived up to its name in proving impossible to achieve. At length, following countless failed attempts, I finally found a way to do it, eight years later.

An idea of the hidden subtleties confronting the would-be designer can be gathered from the following

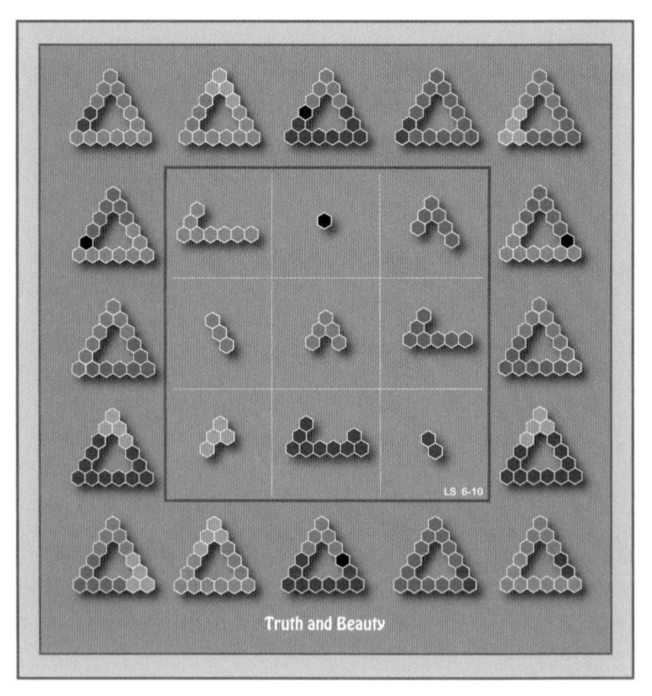

Fig. 24.5 A geomagic square. But is it an interesting case?

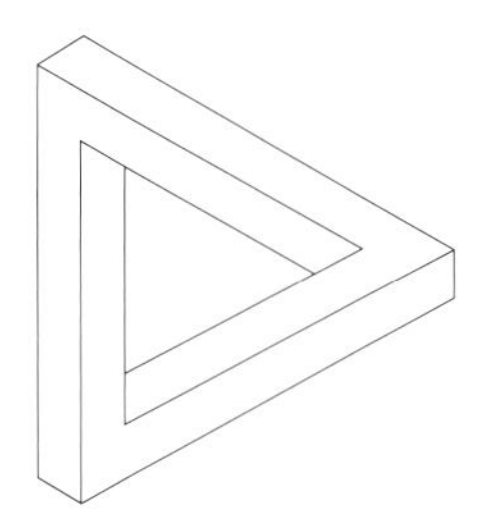

Fig. 24.6 Penrose Tribar, but can it replace the cubes in Figure 24.4?

consideration. It might seem that an equivalent version of Figure 24.4 could be drawn simply by exchanging the pieces-formed-of-cubes by their solid equivalents sawn out from the solid tribar of Figure 24.6. Indeed, this would seem to be so obvious as to hardly require stating. Nevertheless, such an exchange is in fact impossible. Consider, for example, the pair of red cubes in the bottom right-hand cell. From the bottom row target we can see that the length of the equivalent solid beam would be two cubes plus two inter-cube spaces (the latter comprising one whole space between the cubes plus two half-spaces at their ends). The same conclusion is reached on looking at the diagonal target. But now look at the same two cubes in the right-hand-column target. Here, the equivalent length of solid beam will be two cubes plus one and a half spaces, so that our previously sawn solid section will prove too long! The paradox is illustrated in Figure 24.7.

Some might say that Figure 24.4 is not really a geomagic square at all, because the target does not really exist. Nor will it help if the pieces are viewed as two-dimensional, rather than as the 3-*D* objects they suggest. Scrutiny will show that piece outlines in targets often differ slightly from the outline of the same piece in its cell, so that we are not even dealing with a legitimate 2-*D* case.

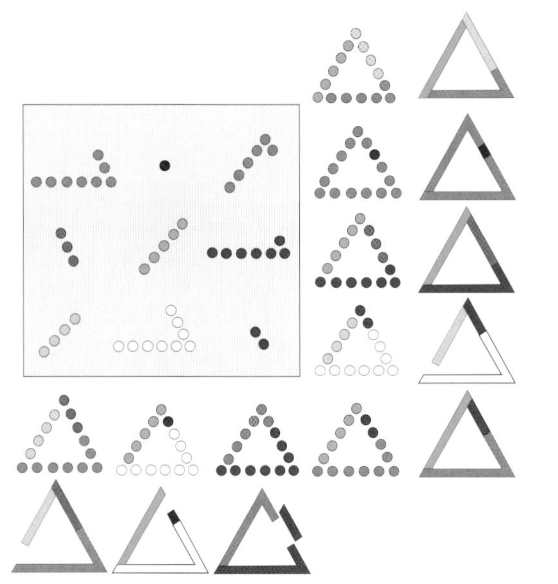

Fig. 24.7 The dots work but the strips don't. Why?

Nevertheless, that a true geomagic square does indeed underlie the tribar square is made evident in Figure 24.5, in which polyhexes replace cubes. Here the title 'Truth and Beauty' was suggested by the compelling logic and simple elegance of the solution.

As described previously in this account, among other enticing prospects competing for attention during the early days of research was the notion of a geomagic square able to tile *two* different targets. A fond dream in a golden mist, as it then seemed. Thereafter we looked at the work of Thoen, de Meulenmeester and Postl, with their amazing 3×3 specimen that surpasses every expectation in achieving the same goal with as many as *twelve* distinct targets. Surely here was a record it would be naïve to try to improve upon. Nevertheless, the square shown in Figure 24.7 not only succeeds in this, it even does so by a comfortable margin.

Figure 24.8 is in fact a 4×4 *semimagic* square, which is to say, a square that is magic along rows and columns only. However, the square has a redeeming factor in that its target may take an infinite range of different forms. Each of the four quadrants is a 2×2 semimagic square showing identical 'half-targets.' The four pieces in each row and column can thus tile any region, connected or disconnected, that is a union of two such half-targets, *irrespective of their spatial relation to each other.* In this way, the possible target shapes are simply unlimited in variety.

For all that, there is no getting around the fact that Figure 24.8 is not fully magic. Moreover, there is not a single piece that is not either weakly-connected or disjoint. Spurred on by these blemishes, the pursuit for perfection

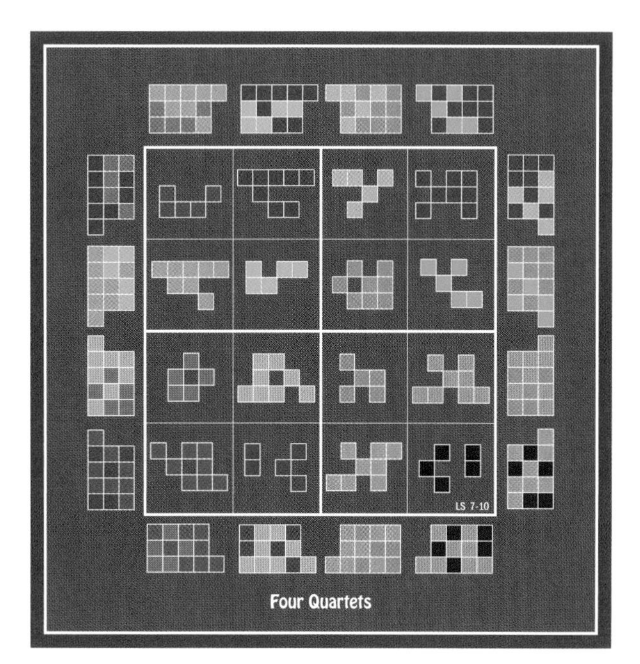

Fig. 24.8 A semi-magic square, but with a redeeming factor.

was rejoined and eventually crowned with success. The 6×6 square in Figure 24.9 surmounts both shortcomings while again enabling an infinite variety of possible target shapes. The latter are again formed by the union of two identical halves. Here, each of the four quadrants is a 3×3 *almost-magic* square (only one magic diagonal), all sharing a common half-target shape.

No discussion of remarkable specimens in this field can overlook two further contributions from the discoverer of the very first order-2 geomagic square, Frank Tinkelenberg. In the first place is his beautiful geomagic star seen in Figure 24.10. Frank ought to change his name to T*w*inkelenberg. It was a prize-winning entry in a competion held by Matthijs Coster in the Dutch periodical *Pythagoras* [15][16]. Note the thought that has gone into his choice of target. Assuming consecutively-sized pieces, its area had to be 33, a number that rules out shapes showing six-fold rotational symmetry. Even so, the hexagonal symmetry of the star is strikingly re-echoed by the hole at the target's centre. Secondly, wielding anew his novel construction principle, originally applied to order-2, Frank then turned to order-3. Figure 24.11 shows his extraordinary find that breaks all previous records for 3×3 magic squares. To begin with, it is a 3×3 panmagic square, which accounts for 12 of the 28 targets shown. The remaining 16 targets are formed by every 4 triads of pieces that can be chosen from any 2×2 subsquare. One seeing is worth a thousand words. Frank tells me that the same nine pieces tile the target in 62 different ways. These nine pieces are a subset chosen from among twelve pieces, using which the target can be formed in 148 different

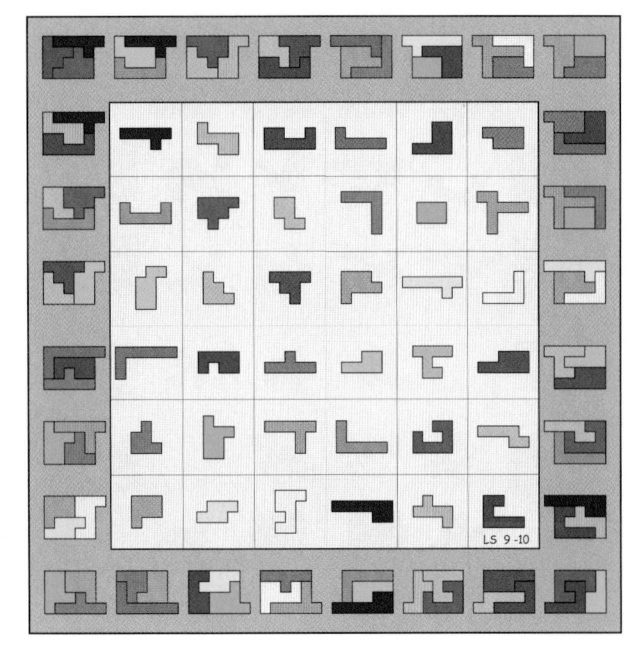

Fig. 24.9 A 6×6 square able to tile an infinite number of distinct targets.

Fig. 24.10 A geomagic star.

Fig. 24.11 A record-breaking 3×3 geomagic square.

ways. My guess is that records will again be shattered once this same technique is applied to 4×4 squares. Hats off to a great achievement!

25 Collinear Collations

The geomagic construction I would like to discuss last is so peculiar as to require a digression.

Martin Gardner once introduced the *Lo shu* with the words, "To appreciate the gem-like beauty of this most ancient of all combinatorial curiosities, consider all the ways that its constant, 15, can be partitioned into a triplet of distinct positive integers. There are exactly eight:

9 + 5 + 1, 9 + 4 + 2, 8 + 6 + 1, 8 + 5 + 2, 8 + 4 + 3,
7 + 6 + 2, 7 + 5 + 3, 6 + 5 + 4." [17]

I can almost hear his subsequent groan. I guess more than one reader must have written to point out his mistake: not "positive integers" but *decimal digits*. However, this doesn't affect the point he was making, which was that these are the self-same set of eight triplets to be found making up the rows, columns, and diagonals of the *Lo shu*. Or, as we might alternatively describe them, every set of three entries *lying in a straight line*. However, the property of 15 he highlighted is not unique. As we saw earlier in connection with Figure 20.8 on page 61, the number 16 can also be partitioned into eight triplets using the same set of distinct digits:

1 + 6 + 9, 1 + 7 + 8, 2 + 5 + 9, 2 + 6 + 8, 3 + 4 + 9, 3 + 5 + 8, 3 + 6 + 7, 4 + 5 + 7,
and so can 14:
1 + 4 + 9, 1 + 5 + 8, 1 + 6 + 7, 2 + 3 + 9, 2 + 4 + 8, 2+ 5 + 7, 3 + 4 + 7, 3 + 5 + 6.

Of course, this doesn't mean that these nine digits can be placed in a 3×3 array so as to yield a magic square with a constant sum of 14 or 16. The centre digit must always belong to *four* distinct triplets: one occupying the central row, one the central column, and two lying along the diagonals. But neither of the above lists contains such a digit. However, with a little ingenuity the same numbers can be placed differently so that eight distinct triplets again

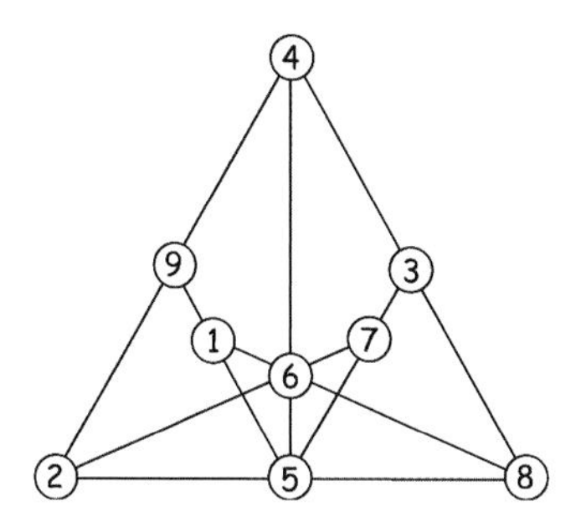

Fig 25.1 The 'Egyptian' *Lo shu*. Every 3 numbers in a straight line sum to 15.

lie along straight line paths. In what I call the 'Egyptian' *Lo shu* seen in Figure 25.1, these triplets again sum to 15:

There are three other ways in which the nine digits can be inscribed in the circles so as to yield a distinct magic pyramid of this kind. Of course, as in the *Lo shu*, the digit 5 again finds itself at the intersection of four straight lines.

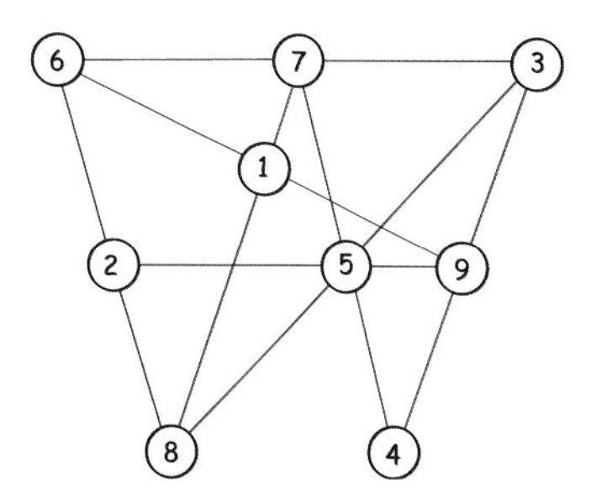

Fig 25.2 The 'Venusian' *Lo shu*. Every 3 numbers in a straight line sum to 16.

But can an arrangement be found such that the magic sum is 14 or 16, rather than 15? Figure 25.2, the 'Venusian' *Lo shu*, arrived at after some trial and error, supplies an answer; its magic sum is 16. Replacing each digit d with $10–d$, results in a similar solution showing a magic sum of 14. It is not difficult to show that for the eight triplets to share the same total then the nine numbers must always form an arithmetic progression.

It was while exploring such systems of collinear triads that I noticed how the nine points could be chosen so as to coincide with the centres of the cells in a 7×7 square array. Soon after, I turned up a similar specimen in which the array was 5×5. Could a yet smaller example be found? A decisive step was to drop the requirement that points need coincide with cell *centres*, so long as they fell *within* cells. In this way, by degrees I arrived at the specimen of 4×4 seen in Figure 25.3, a result that finally jolted me into the realization of what one further step down in size would entail. For the nine points would then occupy all nine cells of a 3×3 array, a result with mischievous potential. With this goal in sight, I thus spared no effort in a diligent search for a solution. On succeeding, the latter then formed the basis for a new puzzle that has been described as 'devilish' [18]. It can be seen in Figure 25.4.

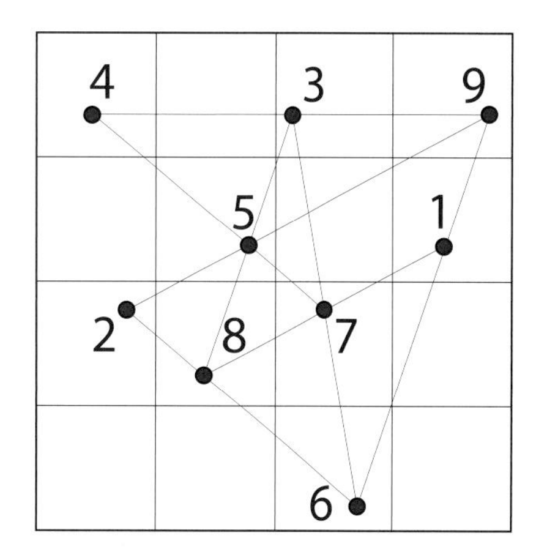

Fig 25.3 Eight collinear triads, the numbered points now falling within the cells of a 4×4 array.

In the diagram below, nine numbered counters occupy the cells of a 3x3 checkerboard so as to form a magic square. Any 3 counters lying in a straight line add up to 15. There are 8 of these collinear triads.

Reposition the counters (again, one to each cell) to yield 8 new collinear triads, but now showing a common sum of 16 rather than 15.

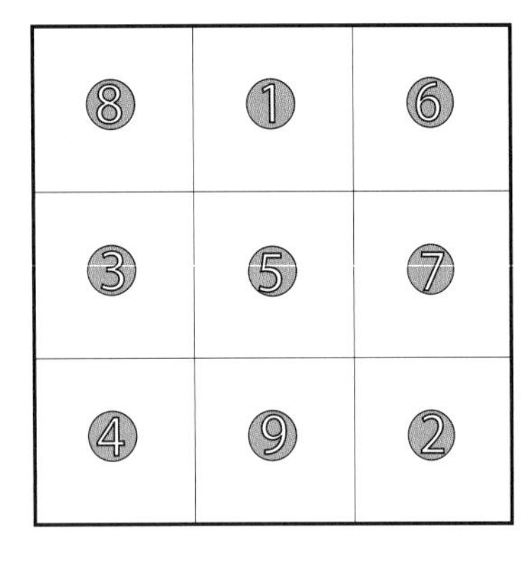

Fig 25.4 A seemingly impossibe challenge?

Fig 25.5 One solution among others.

Fig 25.6 'Hieroglyphs', a symmetrical pattern including 9 collinear triads.

A solution (there are at least four) is shown in Figure 25.5. Alas, the foregoing discussion has already given the game away, so that the challenge posed will seem less impossible here than it would to anyone unfamiliar with collinear structures of the kind discussed beforehand. Among the latter were some readers of *Nieuw Archief voor Wiskunde*, a Dutch mathematical journal in which the puzzle appeared [19], a few of whom submitted 'proofs' of its unsolvability. The latter were, of course, based on their unconscious assumption that counters must occupy the exact centre of each cell. I enjoyed that, naturally.

A point to note is that even the *size* of the counters shown may not exceed a certain limit if they are to remain within their cells. An alternative puzzle that presents a similar picture of numbered counters poses the different challenge of rearranging them so as to form 8 new collinear triads, again each summing to 15, but with 1 now occupying a corner cell. Readers who may like to seek a solution will find that it can be done using counters that are a good bit bigger than those that appear in the above puzzle.

The collinear point clusters we have been looking at may have their attractions, but can hardly be said to have visual appeal. The examples here shown are but a few among a large variety of solutions discovered, but none of them especially pleasing to the eye in being symmetrical or elegant, save in one outstanding case. It is also the sole instance known to me of a set of nine points that include *nine* rather than eight collinear triads. It is 'Hieroglyphs,' seen in Figure 25.6, a diagram showing three-fold rotational symmetry, but in which pentominoes appear rather than numbers. There are two reasons for this replacement. In the first place, it is not difficult to show that an equivalent structure using nine distinct numbers is impossible to construct. In the second, a switch to planar forms brings with it the usual increase in the number of targets that can be tiled. Thus it is that besides the nine collinear triads, all six triads of cells arranged in an equilateral triangle formation are magic also. These fifteen targets can be seen at the base of the pyramid. Note the recourse to nocturnal photography in capturing this specimen.

26 Concluding Remarks

If it strikes the reader that, in the previous section, we have strayed somewhat from the path of magic squares and begun to explore too far afield, I can only own up. It is a sign that the end is in sight. Much like the traveller returned from a journey in foreign lands, in the foregoing pages I have illustrated the tale of my explorations with souvenirs and specimens brought back from the expedition. But such specimens are not unlimited in supply. At least, not if repetition is to be avoided. Time thus to bring this account to a close before such an accusation can be levelled. I conclude with a few parting observations.

It may have been noticed that, with a single exception, magic squares larger than 4×4 have escaped attention in the foregoing account. The reason is simply that larger squares have seldom struck me as attractive. How many readers are going to be tempted to add up swarms of numbers in order to verify that a square is magic? A dedicated few only, I cannot help thinking. And what then of the *elegance* that is the very hallmark of the (small) magic square? Similar reservations apply to non-numerical squares. Moreover, it is a common fallacy that the bigger the square, the greater the achievement it represents, because of the supposed difficulty of getting so many numbers to comply with the magic conditions. But this idea is quite simply mistaken. It is easily shown that the constraints imposed in fact *diminish* with increasing size of square. For, as Chernick [20] was the first to prove, an algebraic generalization of an $N \times N$ numerical magic square can always be written so that $N^2 - 2N$ of its cells are each occupied by a single independent variable. Or in layman's terms: in creating a magic square of order-N, the proportion of cells that can be assigned *any* desired number is $1 - 2/N$, a proportion that increases with N, and is already as much as *two-thirds* of the cells for a square of size 6×6.

These ruminations are a reminder of a related question. I refer to the problem of how the 'magicality' of a square is to be assessed. The question receives next to no consideration in the literature, and yet the matter of what makes one magic square more magic, and hence more praiseworthy than another, is really of central importance to the whole topic. Magic-square buffs are at heart trophy-hunters, the essence of the game being to track down new specimens that outclass the opponent's bag. But how, in the absence of yardsticks or scales, are meaningful comparisons to be made among different exhibits? That there is no easy answer to the question is to be seen by considering the case of that most famous square of all, the universally acclaimed *Lo shu*.

What do we find when we look at the numbers appearing in the *Lo shu* when pictured as a collection of points distributed along the real number line? The answer is shown in Figure 26.1. It is a set of nine equidistant points positioned slightly to the right of zero, which is at the origin, or centre of the line.

But what kind of an aesthetically insensitive soul, I wonder, has left this peerless string of pearls lying carelessly in such an *asymmetrical* position? And why has the right- hand side of the line been chosen over the left? Is this not blatant *signism*, a form of discrimination against negative numbers?

Oh come now, I hear a critic rejoin, what possible difference can the *position* of the nine points make? Their key property is that they form an arithmetic progression, which is to say, they form a chain of equidistant links. Placing them where they are yields us a magic square that contains the first nine counting numbers, which are

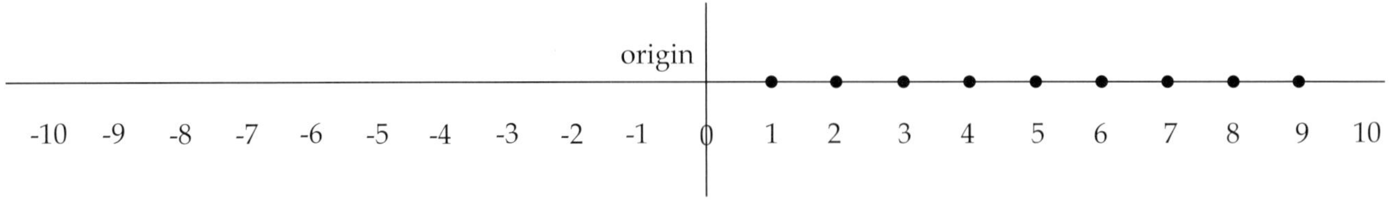

Fig. 26.1 A string of pearls.

-3	4	-1
2	0	-2
1	-4	3

2	9	4
7	5	3
6	1	8

Fig. 26.2 The balanced square and its close relative the *Lo shu*.

simple, pleasing, and immediately recognized by all. Lay them anywhere else and the result will be an essentially identical magic square but one lacking in these desirable qualities. And in any case, can any better position on the line be found?

It can. When it comes to magic squares, symmetry and magicality are the two sides of a single coin. Increase one and the other will be augmented. Decrease one and the other will diminish. As we shall see, the asymmetrical position of the points used in the *Lo shu* exacts a price that is paid for through a relative *reduction* in its magical properties.

To see this, let us slide our set of equidistant points leftwards along the line to the only position of true balance or symmetry, which is in the exact centre. The central number of the progression now coincides with zero, and the result is the 'balanced' magic square of Figure 26.2 (left), whose simplicity may belie its enhanced magicality. For comparison, the *Lo shu* appears at right.

To begin with, consider the sum of the four corner numbers in the *Lo shu*. It is $2 + 4 + 6 + 8 = 20$. In our new square it is $-3 + -1 + 1 + 3 = 0$, the same value as the centre number. Or in other words, the centre number is equal to the sum of its four immediate diagonal neighbours. But now consider both squares as if *toroidally-connected*. The four diagonal neighbours of any cell are then the same as they would be were the square used as a tile with which to cover the plane. This brings to light an arresting fact. For, unlike the *Lo shu*, *every* entry in the left-hand square is now found to be equal to the sum of its four diagonal neighbours.

Now is that not a 'magical' property to celebrate? Had the same thing been true of the *Lo shu*, would it not have been trumpeted abroad at every opportunity? However, there is more to come.

Call the product of the numbers in any row or column of a magic square, the row-product or column-product. Now Hahn[21] has shown that a property enjoyed by every 3×3 magic square is that the sum of its three row products is equal to the sum of its three column products[8]. In the *Lo shu*, for example, this sum of products is 225.

It would be pleasing were a similar property to be shared by the three \-diagonals or 'slopes,' and the three /-diagonals or 'slants,' but in fact it is not. Not in every case, that is. Not in the *Lo shu*, in particular. But it *is* true in the balanced square, where the sum of the 3 row-products is equal to the sum of the 3 column-products *and* the sum of the 3 slope-products is equal to the sum of the 3 slant-products:

$$(-3 \times 4 \times -1) + (2 \times 0 \times -2) + (1 \times -4 \times 3) = (1 \times -3 \times 2) + (4 \times 0 \times -4) + (-1 \times -2 \times 3) = 0$$

and

$$(4 \times -2 \times 1) + (-3 \times 0 \times 3) + (-1 \times 2 \times -4) = (4 \times 2 \times 3) + (-1 \times 0 \times 1) + (-3 \times -2 \times -4) = 0.$$

Moreover, the sums of the row-products, column-products, slope-products and slant products are all the same; they are all equal to the magic sum, zero.

Here then is another 'magical' property of the balanced square that is lacking in the *Lo shu*. How come it is absent from the latter? A glance at any generalization of order-3 squares is sufficient to explain all. From this will be seen that the sum of the four corner entries in any 3×3 magic square is always equal to four times the centre number. But $4x = x$ only when $x = 0$. That is, the four corner entries will sum to the centre entry only if the nine numbers employed are *centered on the origin of the real number line*. It is as foretold: enhanced magicality goes hand in hand with enhanced symmetry.

It is a paradoxical fact that although much of the *Lo shu*'s appeal lies exactly in its use of the natural numbers, these are now seen to be the source of its lesser magicality when compared with the square above. Not that I suppose these

---

8 Curiously, Hahn failed to notice an important point. The property he identified belongs, in fact, to every 3×3 semimagic square. It is only because fully magic squares are a particular instance of the latter that they inherit the same property.

assaults on the *Lo shu* will do its reputation any lasting harm. However, if my example has helped to illuminate the difficulties involved in comparing the magicality of different squares then it will have served its purpose.

The generalization of order-3 squares just mentioned brings me nicely to a final remark. It will come as no surprise to anyone who has read this book that I had a particular generalization in mind.

Édouard Lucas's formula has been a source of fascination to me since the day I first beheld it and saw he had improved on a similar square I had arrived at on my own. Admittedly there is not much to choose between the two (seen in Figure 26.3), which are merely alternative expressions of exactly the same algebraic generalization.

$c+a$	$c-a-b$	$c+b$
$c-a+b$	c	$c+a-b$
$c-b$	$c+a+b$	$c-a$

Lucas

$c+a$	$c+b$	$c-a+b$
$c-2a-b$	c	$c+2a+b$
$c+a+b$	$c-b$	$c-a$

author

Fig. 26.3

But Lucas's formula uses 33 symbols, whereas mine requires 35. And while his does it without requiring numbers, mine employs two 2's. I guess many mathematicians will have little patience with such pernickety attachment to appearances, a standpoint I can understand and respect, but which is simply not my style, who feel more of a *poet* than mathematician. And on grounds of economy of expression there is no denying that of the two 'poems' in Figure 26.3, Lucas's is the better. By way of comparison, Figures 26.4 and 26.5 show further versions of the same generalization, the first taken from an article by Martin Gardner, the second, which is unusual in showing that the three numbers in the top row may be chosen at will, again due to the author.

$a+x$	$a+2y+2x$	$a+y$
$a+2y$	$a+y+x$	$a+2x$
$a+y+2x$	a	$a+2y+x$

Fig. 26.4

a	b	c
$\frac{-2a+b+4c}{3}$	$\frac{a+b+c}{3}$	$\frac{4a+b-2c}{3}$
$\frac{2a+2b-c}{3}$	$\frac{2a-b+2c}{3}$	$\frac{-a+2b+2c}{3}$

Fig. 26.5

As related earlier in this volume, it was the desire to come up with a pictorial representation of Lucas's formula that first lead to the idea of geometric magic squares. Even before that, however, the feeling that something could be "done" with Lucas's square had haunted me for years. In my own mind it was not merely a kind of poem, but an almost mystical object that enshrouded some potent secret, like the green jewel in the forehead of a golden idol. At the back of my mind was always a vague idea that its algebraic symbols could be translated into some sort of graphical equivalent that would make this secret manifest. A kind of vindication of this intuition came in 1997, when I tried substituting Gaussian integers for variables and thus produced magic squares whose nine entries could be plotted as points on the complex plane. The discovery that these nine points always lay on the periphery of a parallelogram could be formulated as a new theorem, elementary, yet previously unknown. If there was one area of magic square theory in which one would least expect to make such an interesting find it was surely that of 3×3 squares, the smallest, and hence most thoroughly investigated type of all, with a history going back over some 2,000 years. The finding appeared as "The Lost Theorem" published in *The Mathematical Intelligencer*. Its appearance engendered warm compliments from a couple of readers, but I'm afraid the truth is that it has since passed into oblivion, so that the Parallelogram Theorem of 3×3 magic squares remains as unknown today as it was before I discovered it. The piece has been included here as Appendix 5 in the hope of reviving its memory. In similar fashion, it is a mystery to me why Lucas's square is not celebrated as the greatest of all 3×3 magic squares, embracing as it does all the myriads of its particular numerical instances, whilst enshrining the timeless symmetries that govern the structure of them all.

Appendix I

A Formal Definition of Geomagic Squares

An informal definition of geomagic squares has been given elsewhere in this book. From the mathematical point of view, however, loose talk about 'pieces fitting together as in a jigsaw' is clearly inadequate. The purpose of this addendum is thus to provide a formal definition of geomagic squares. To this end, we begin by defining a 'generic' magic square, by which is intended a generalized magic square in its widest possible sense. That is, a canonical magic square is described, but without stipulating the kind of 'objects' that occupy its cells, and without specifying how these combine so as to form a constant target ('target' again in the broadest sense possible).

Although the presentation has been kept as informal as possible, a degree of formality is necessary. No doubt some mathematicians will find the treatment too lax; for laymen it may prove too demanding. At the same time, certain details have inevitably escaped mention, just as some of the mathematical concepts involved could have been explained more fully.

Properties of the Generic Magic Square

Fundamental to any conceivable magic square (of order $n \geq 2$) is an $n \times n$ matrix M over a domain D, in which we discern $2n + 2$ (potentially 'magic') lines: the n rows, the n columns, the main diagonal (\) and the co-diagonal (/).

All essential properties of the generic square of order $n \geq 2$ can now be specified in a list of 5 components:

i) A domain D, which is a set of 'objects' that will occupy the cells of M. In the case of numerical magic squares, this domain would be simply the ring of integers Z.

ii) A binary operation + on D. Since the n objects in every line must be able to merge so as to form the same 'sum' in each case, we require that D be equipped with a rule of combination. Our requirements will be met by a binary operation + that is:

commutative ($d_1 + d_2 = d_2 + d_1$),
associative $d_1 + (d_2 + d_3) = (d_1 + d_2) + d_3$,
and has an identity element '0,' such that
$0 + d = d + 0 = d$ for all $d \in D$.
We shall use the term 'sum' for the operation +. For $D = Z$ the binary operation is ordinary addition and the identity element is zero.

iii) An equivalence relation ~ on D. For $D = Z$ this equivalence relation is equality ('='). Note that compatibility of the equivalence relation with the binary operation is not required. This means that for a, b, a', $b' \in D$ with $a \sim a'$ and $b \sim b'$, it is not necessarily true that $a + b \sim a' + b'$.

iv) An element $T \in D$, called the 'target' or 'magic sum'. For $D = Z$, T is thus an integer – the magic sum.

v) Two rules that the matrix M must obey. These are:

1. For each line L, the following Line-Sum Condition holds:

<u>LSC</u> :

Suppose $L = \{d_1, \ldots, d_n\}$. Then there exist elements $d_1', \ldots, d_n' \in D$ such that $d_i \sim d_i'$ for all $1 \leq i \leq n$ and $d_1' + \ldots + d_n' = T$.

This means that the sum of $d_1, \ldots, d_n$ is not necessarily equal to the target but that there exist elements $d_1', \ldots, d_n'$ which *do* sum to the target and are equivalent to $d_1, \ldots, d_n$ respectively.

2. The n^2 elements in the matrix M are mutually non-equivalent.

Rule 1 guarantees that the square will be *magic*, which is to say, that the n elements in each line combine to produce the same outcome (the target). Rule 2 ensures that the square is not trivial, which is a term used to describe squares showing repeated entries. An $n \times n$ matrix M over a domain D satisfying rules 1 and 2 is thus a *non-trivial magic square* of order n.

Remark

It is important to understand that the content of a cell of the matrix M (i.e an element of the domain D) is a representative of an equivalence class (as defined by ' ~ '), which is to say, a set of equivalent objects. For example, a cell may contain a planar piece of a certain shape, whereas that particular piece is merely one member of an equivalence class consisting of the 'same' piece in all its myriad possible alternative positions and orientations. Note that this makes no difference in the case of numerical magic squares because equivalence is then simply equality, so that each equivalence class consists of a single number. But for geometric magic squares an equivalence relation is essential because although two identically shaped pieces in the plane, say, are not equal, they are congruent, which for our purpose means equivalent.

With the generic square now described, we can turn to the definition of geomagic squares, which are a particular instance of the generic square. Accordingly, our goal will be achieved by providing a suitably precise specification of each of the five components.

The Domain (i)

By way of introduction to the domain to be used, consider closed, bounded subsets of the Euclidean space of dimension d ($d \geq 1$). A subset C is 'closed' if and only if all boundary points of C belong to C. An example is the unit disk, which consists of all points of the plane having distance ≤ 1 from the origin. It is a closed set because its boundary, the unit circle, belongs entirely to it. If we remove any number of points from the boundary (the unit circle) then the resulting set is no longer closed.

Further, a subset B is 'bounded' if and only if there exists a sphere of sufficiently large radius such that B is completely surrounded by it. In other words, B does not 'extend to infinity' in any direction. Subsets of Euclidean spaces, which are both closed and bounded are known as 'compact' subsets. In fact, in general topology, the definition of 'compact' is different, but in the special case of Euclidean spaces the two definitions are equivalent, so that 'compact = closed + bounded', as shown by the Heine-Borel theorem. In light of these remarks, our domain D will be the set of all compact subsets of Euclidean space of dimension d ($d \geq 1$). Hence, by a geomagic square of dimension d will be meant one in which every entry is a compact subset of the Euclidean space of dimension d.

The remaining 4 components can now be identified as follows:

The Binary Operation (ii)

For two compact subsets A and B of the space, let $A + B := A \cup B$ (set union). The identity element 0 is then the empty set Ø.

The Equivalence Relation (iii)

For two compact subsets A and B of the space, let $A \sim B$ if and only if A and B are *congruent*, that is, there exists an invertible, affine linear distance-preserving transformation h of the space such that $h(A) = B$.

This means that the set B can be obtained from the set A by a displacement of A through space, involving rotations and/or translations, together with a possible reflection.

The Target (iv)

The target T is any compact subset of the space, that is, a member of D.

The Line-Sum Condition (v)

The two rules to be obeyed by the matrix M remain unchanged for geomagic squares, although the latter do impose an additional requirement on the Line-Sum Condition:

The elements $d_1', \ldots, d_n'$ (see LSC) are allowed to touch each other, but not to overlap. This means that for all $i \neq j$ the intersection of d_i' with d_j' is empty, or consists of common boundary points of d_i' and d_j' only. That is, the intersection contains no interior points.

Expressed informally, the Line-Sum Condition for geomagics thus demands that the elements of any line L can be arranged so as to *pave* the target, without overlaps.

This completes the definition of geomagic squares.

A Note on Numerical Vs Geomagic Squares

In the case of dimension 1, the compact, *connected* subsets are the closed, bounded intervals of the real number line, such as the interval [0,1], meaning real numbers in the range 0…1. Numerical magic squares using real numbers can then be viewed as geomagic squares for which (i) the domain D is the set of all bounded, closed intervals of the real number line, and (ii) the target is again an interval. The binary operation + and the equivalence relation ~ are then set union and congruence, and since two bounded, closed intervals are congruent if and only if they are of the same length, numerical magic squares over the reals are revealed as a special case of geometric magic squares of dimension one. They are that case in which the compact subsets are all *connected*.

A further point of interest is that numerical magic squares operate under constraints that do not apply to geomagic squares when $d > 1$. For example, no numerical 3×3 panmagic squares exist, whereas for $d = 2$, an example can be found on page 17. Moreover, this same square even exhibits a further four symmetrically arranged magic triads that enhance its magicality even beyond that of a panmagic square. Whence comes this greater potency enjoyed by geomagic squares?

Suppose we have $a + b = c$, where a, b and c are real numbers. From this, we know that b is the only number, that can be added to a so as to yield c. And the same obtains if a, b and c are three closed, bounded intervals of the real number line, rather than numbers.

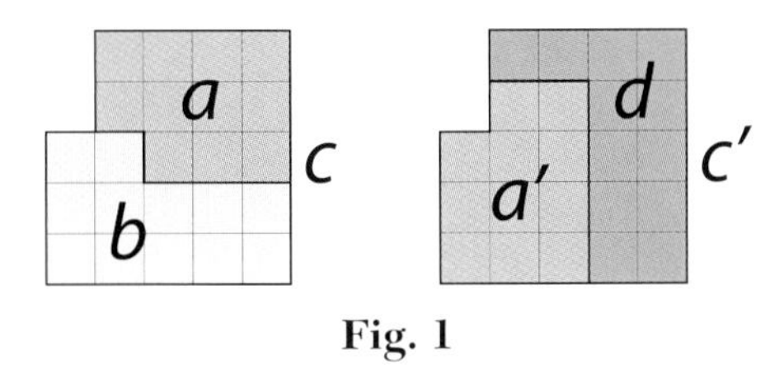

Fig. 1

But suppose now that a, b, c (and d) are compact subsets of the Euclidean space of dimension 2 corresponding to the planar pieces seen in Figure 1. Note that $a' \sim a$ and $c' \sim c$. But here b is no longer the only piece that can be appended to a so as to yield c. An alternative is the different piece d, because $a + b \sim a' + d \sim c' \sim c$. In short, the behaviour of compact subsets is inherently more flexible than is that of numbers.

Geomagics enjoy more freedom than numerical squares because their targets can be assembled from members of their domain in many more ways than magic sums can be formed with numbers.

We have seen that numerical magic squares are a subset of geomagic squares. Could it be that geomagic squares are themselves a subset of some wider genus of magic square? The question remains unanswered. In any case, the plethora of new challenges thrown up by 2-D geomagic squares ought to prove sufficient to keep investigators occupied for some time to come.

Grateful thanks are due to my friend Michael Schweitzer who provided the mathematical basis of the above definition.

Appendix II
Magic Formulae

Introductory Note

Penned in 1980, but never published, 'Magic Formulae' was my first ever attempt at writing up a new result in the form of an essay. Knowing little at that time of the world of professional journals, the style adopted is unusual, my imaginary audience being composed of laymen possessed of a smattering of mathematics, such as myself. Thoughts of publication not having been seriously entertained, I saw myself as my own publisher, and thus took a good deal of trouble over the presentation, the completed article taking shape as an A5 booklet sandwiched between card covers.

We didn't have word-processors in those days, so the whole thing was bashed out on my old Adler typewriter, matrices and all. For the front and back covers, I made drawings of algebraic magic squares in which the variables were represented by 'magical' symbols: runes and astrological signs. The letters used for the title 'Magic Formulae' were copied from the Book of Kells. In retrospect I can see that this concern with symbology marked the beginning of a train of thought that would chug along for some thirty years until its eventual arrival at geometric magic squares. Should anyone be interested in how the author's mind works, here then is a good place to start. With the present volume in mind, in the interests of clarity I recently returned to the piece and improved it in various respects.

Algebraic generalizations using runic characters and astrological signs embellished the front and back covers of Magic Formulae. Red symbols stand for positive variables, black symbols for negative variables. The square at left generalizes panmagics, while the other covers all 4x4 magic squares.

Magic Formulae

"Runes and charms are very practical formulae designed to procude definite results, such as getting a cow out of a bog."
—T.S. Eliot, The Music of Poetry

In the year 1910, Ernest Bergholt, a familiar name in the British literature on recreational mathematics of the time, published a new generalization of 4×4 magic squares [1]. He gives us no clue as to how he arrived at it, but simply presents it with the claim that ". . . it is the completely general formula . . ."

$A-a$	$C+a+c$	$B+b-c$	$D-b$
$D+a-d$	B	C	$A-a+d$
$C-b+d$	A	D	$B+b-d$
$B+b$	$D-a-c$	$A-b+c$	$C+a$

Fig. 1 Bergholt's generalization.

Bergholt uses the word 'formula' in the sense of a recipe, and by a completely general formula he means one that generalizes all possible 4×4 magic squares, rather than some restricted family of special types. Below we shall sometimes use the term *universal* generalization to indicate the same thing. Note how the formula neatly encapsulates many intrinsic properties of every 4×4 magic square, such as that the four corner numbers, the four central numbers, the four central numbers in the outer rows, and the four central numbers in the outer columns all sum to the same total as do the rows, columns and diagonals: $A+B+C+D$. Alternative proofs of these properties, often tedious and space-consuming, nevertheless clutter the literature, showing that the advantages of such algebraic formulae remain anything but widely recognized.

H. E. Dudeney, the famous British puzzlist, was the first to acknowledge significance in Bergholt's square. It is ". . . of the greatest importance to students of this subject," he says in *Amusements in Mathematics* [2]. Much later, H. S. M. Coxeter found it sufficiently interesting to include in the eleventh edition of *Mathematical Recreations and Essays* [3]. Only three years after this, however, even Maurice Kraitchik, who devoted several pages of his *Mathematical Recreations* to algebraic formulae, makes no mention of it. Since then Bergholt's contribution seems to have been all but forgotten.

$c+a$	$c-a-b$	$c+b$
$c-a+b$	c	$c+a-b$
$c-b$	$c+a+b$	$c-a$

Fig. 2 Lucas' 3×3 formula.

Only for a brief moment in 1938 did Bergholt's square surface from obscurity and threaten to claim a permanent place for itself in the theory of magic squares. In that year, Jack Chernick published a method for constructing generalizations of any order in a standard normal form [4]. But through a circumstantial quirk, Chernick's technique was applicable only to 5×5 and larger squares, so that for the sake of completeness he was obliged to include non-standard generalizations for orders 3 and 4. For the former, he used the widely known formula due to Édouard Lucas [5] seen in Figure 2.

When it came to order-4 however, Chernick departed from his most obvious course, which was to use a non-standard, but still Chernick-like formula such as that seen in Figure 3, and reproduced instead Bergholt's square. This suggests that he appreciated which of the two was the better. Bergholt's square is not only more economical in expression, it reveals that a *Latin* square is concealed within every 4×4 magic square, something we may never have guessed from Figure 3. [By a Latin square of order n is meant one showing n^2 entries of n different elements, none of them occurring twice within any row or column, as seen in the arrangement of upper-case letters in

Bergholt's formula]. I suspect that this may have secretly irritated Chernick, because, instead of enthusing over Bergholt's gem, he went on to quibble over a trifling point in the proof of its generality. What is amusing is that even after Chernick had finished tinkering and had satisfied himself that all was well, Bergholt's square remained *unproven*.

The criteria of proof implied by Bergholt's paper in support of the complete generality of his square were threefold:

1) the square is magic
2) every cell is occupied by a distinct entry
3) it is a function of 8 independent variables.

Following an interlude of twenty-eight years (research in this area proceeds at a leisurely pace), Chernick protested that the formulation of the third criterion had

p	q	r	s
t	u	v	$p+q+r+s-t-u-v$
w	$p+t+w-s-v$	$q+r+2s-t-u-w$	$u+v-w$
$q+r+s-t-w$	$r+2s+v-t-u-w$	$a+e+f+h-c-d-g$	$t+w-s$

Fig. 3 A Chernick-type generalization of order-4.

To create the universal generalization of 4×4 magic squares in Figure 3.

(Below cells are indicated as in the adjacent square; k is the magic constant.)

A	B	C	D
E	F	G	H
I	J	K	L
M	N	O	P

To create Figure 3 from the adjacent square,

Let $A = p;\ B = q;\ C = r;\ D = s;$
Hence $k = p + q + r + s$ by definition
Let $E = t;\ F = u;\ G = v;$
Hence $H = k - E - F - G = p + q + r + s - t - u - v$
(orthogonals and diagonals sum to k)
Let $I = w;$
Hence $M = k - A - E - I = q + r + s - t - w$
and $J = k - D - G - M = p + t + w - v - s$
and $N = k - B - F - J = r + 2s + v - t - u - w$
Now, let $K = x;$
Hence $O = k - C - G - x$
and $P = k - A - F - x$
but $L = k - D - H - P = k - I - J - x$
from which $x = D + H + P - I - J = q + r + 2s - t - u - w$
Substituting for x, we get

$O = p + t + u + w - r - s - v$
$K = q + r + 2s - t - u - w$
$P = t + w - s$
$L = u + v - w$ which completes the square.

Since $p,q,..$ represent any numbers, and the only limitation placed upon other entries is satisfaction of the necessary and sufficient conditions for magicality, the result is a universal generalization of 4×4 magic squares.

been "stated without proof." Nevertheless, he went on to show that Bergholt's figure of 8 had indeed been correct, and advanced even further by proving that any universal generalization of order-n magic squares is necessarily a function of $n^2 - 2n$ independent variables. In the case of order-4, this figure is of course $4^2 - (2 \times 4) = 8$. So, not only had Chernick vindicated Bergolt's formula, his inclusion of it in his own paper ought to have lent it a degree of prestige. Nevertheless, during the ensuing forty years Bergholt's work has gradually settled further into oblivion, and here the matter has rested until now.

Chernick's elementary procedure for arriving at Figure 3 is detailed in the box below. It is important to realize that, in contrast to Bergholt's approach, it is this very process by which Chernick formulae are arrived at that is a guarantee of their validity. This remains the case whether the formula derived be universal in scope or otherwise.

Returning now to Bergholt's work, Figure 4 shows a matrix which, although satisfying all three of his conditions, still fails to be a universal generalization. Figure 4 is in fact a *Graeco-Latin* square, which is to say, one formed by the addition of two Latin squares so as to yield a distinct entry in each cell. Moreover, the latter are *diagonal* Latins, or ones whose 4 distinct elements also appear in the two main diagonals. But were Figure 4 also a generalization of all magic squares of order-4, the four cells forming each quadrant would *not* sum to the magic constant, $A + B + C + D + a + b + c + d$, as they do here. This is a property of *some* 4×4 magic squares, but certainly not of all. Bergholt's criteria are, therefore, revealed as flawed, since they fail to distinguish been universal and non-universal generalizations. But then Bergholt's generalization itself becomes open to suspicion, the question of its true generality once again coming into question.

$A+a$	$B+b$	$C+c$	$D+d$
$C+d$	$D+c$	$A+b$	$B+a$
$D+b$	$C+a$	$B+d$	$A+c$
$B+c$	$A+d$	$D+a$	$C+b$

Fig. 4 A diagonal Graeco-Latin square.

Not surprisingly, these doubts were far from obvious to me on my first encounter with Bergholt's formula. Like others before, I accepted his square without question. Yet the simplicity and almost palindromic symmetry of his result intrigued me so much that I began to speculate about how he had found it, as well as whether he had extended his approach to higher orders. A search of the literature turned up no references however, so that I began in earnest on an attempt to create a similar generalization for order 5. Eventually my efforts were rewarded, and I am pleased to present here the hoped-for order-5 square; see Figure 6. However, a very surprising development has been the discovery of a further simplification of Bergholt's square, as shown in Figure 5. Proofs that Figures 5 and 6 are indeed universal generalizations can be found in the boxes below.

A	$B+a$	$C+b$	$D+c$
$C+c+x$	$D+b$	$A+a$	$B-x$
$D+a-x$	C	$B+c$	$A+b+x$
$B+b$	$A+c$	D	$C+a$

Fig. 5 A minimal formula for order-4.

Admittedly, the simplification of Bergholt's square is modest, an application of Occam's razor resulting in little more than a close shave. On the other hand, it is easy to see that the eight variables occurring in any order-4 formula must each appear at least four times: one for each row/column when the variable appears in the magic constant; twice positive and twice negative when not. The minimum number of variable appearances is thus $8 \times 4 = 32$, as in Figure 5, for which reason we may call this a *minimal* formula.

Proof that Figure 5 is a universal generalization of order-4

In Figure 5, let

$A = p$
$B + a = q$
$C + b = r$
$D + c = s$
$C + c + x = t$
$D + b = u$
$A + a = v$
$D + a - x = w$

From these equations we can derive:

$A = p$
$B = p + q - v$
$C = p + t + w - s - v$
$D = p + t + u + w - r - s - v$
$a = v - p$
$b = r + s + v - v - t - w$
$c = r + 2s + v - p - t - u - w$
$x = t + u - r - s$

Substituting $p, q, \ldots$ for $A, B, \ldots$ in Figure 5, we derive a Chernick-type generalization (Figure 3); the two squares are thus isomorphic.

$A+a$ $+x$	$B+b$ $-z$	$C+c$ $-v-x+z$	D	$E+d$ $+v$
C	$D+d$ $-x$	$E+a$ $+v+x+y$	$A+b$ $-v$	$B+c$ $-y$
$E+b$ $-u-x-y$	$A+c$ $+u+x+z$	B	$C+d$ $+v+w-z$	$D+a$ $-v=w+y$
$B+d$ $+y$	$C+a$ $-u$	$D+b$ $+u+w-y$	$E+c$ $-w$	A
$D+c$ $+u$	E	$A+d$ $-u-w-z$	$B+a$ $+z$	$C+b$ $+w$

Fig. 6

Figure 6 is in fact not a minimal formula, the very first example of which being due (to the best of my knowledge) to Prof. Don Knuth, with whom I corresponded on this topic. Later, a puzzle based on the problem of distinguishing universal from non-universal formulae appeared under our names in the *Journal of Recreational Mathematics*[1]. Knuth's minimal formula which featured in the puzzle presentation is shown in Figure 7.

The foundation of Bergholt's approach rests in the observation that a numerical diagonal Latin square is simultaneously a trivial type of magic square in which only n of the n^2 entries are distinct; whereas a literal diagonal Latin square (i.e., one using letters) can be interpreted as an algebraic generalization of such trivial squares. He then had the ingenious idea of introducing modifications to the literal Latin square in a way that would preserve its magic properties while ensuring that each cell becomes occupied by a distinct entry. This is effected by the arrangement of lower-case letters in his matrix. Similarly, both of my own generalizations above are elaborations upon *Graeco-Latin* squares. But Bergholt was still up against the problem of deciding whether a matrix so produced is a generalization of *all* 4×4 magic squares or not. For there is no guarantee that the relations established among the cells are not restrictive *beyond* the point dictated by the definition of a magic square. Having grasped his principle, it was not difficult to dream up candidate squares for order-5, but the question remained of how they were to be tested. The solution to this problem is of course pivotal in identifying genuinely universal formulae from among likely contenders.

Following a lot of thought, eventually I came to see that if Chernick's generalization was by definition universal (see first box above), any alternative formula could only be a re-expression of that same generalization in different form. The test of a putative formula, in short, stood or fell by a demonstration of its *isomorphism* with the Chernick-type square. Two of the above boxes provide demonstrations of isomorphism between Figures 5 and 6 and their corresponding Chernick equivalents. It is ironic and fitting that Chernick held the key to Bergholt's vindication all along: for Bergholt's formula is indeed a "completely general formula" as may easily be verified in the same way as shown for Figure 5.

Reducing Formulae to Essentials

The formulae looked at above have all been generalizations of traditional or *additive* magic squares, which is to say, those in which the sums of the numbers in every row, column, and diagonal are alike. Analogous formulae for *multiplicaitive* squares, showing constant products, can equally be created. On refelection, however, we see that, all other things being equal, the only change required in any such matrix would be the replacement of × for +, and 1/a for $-a$; where a may be any variable. This realization quite naturally expands to the unspecified binary operation,~, and the concept of a generalization of "~-itive" magic squares, similar in other respects to additive types, save that ~ replaces +, and the inverse a^{-1} replaces $-a$. Still further compression can be achieved by writing ab for $a \sim b$, and $\overline{a}$ for $-a$. These points find

A	$B+u-v$	$C+c+v-w$	$D+d-u+w$	$E+e$
$D+c+x-y$	$E+d$	$A+e-x$	B	$C+y$
$B+e+y-z$	$C-u$	D	$E+c+u+b$	$A-b+d-y+z$
$E-x+z$	$A+c$	$B+b+d+x$	$C-b+e$	$D-z$
$C+d$	$D+e+v$	$E-b-v+w$	$A-w$	$B+b+c$

Fig. 7

1 Problem 1296 JRM Vol 16 No 2 1983-4 p 138

expression in Figure 8, a square that subsumes additive and multiplicative magic squares in a single formula. By discarding redundant operator signs, we not only increase the sweep of generalizations, but expose the essential structure with starker clarity.

A	Bay	$Cb\overline{y}$	c
Ccx	b	Aa	$B\overline{x}$
$a\overline{x}$	C	Bc	Abx
Bb	$Ac\overline{y}$	y	Ca

Fig. 8 Another minimal formula for order-4.

This reduced form will be adopted in what follows, although in the discussion it will be convenient to retain the familiar terminology of additive magic squares. Bear in mind though that justaposition (*ab*) will represent any binary operation, and $\overline{a}$ the inverse of a, so that the scope of variables is not restricted to numbers only, but embraces the elements of any abelian group. Figure 8, incidentally, is a second example of a minimal formula based on a diagonal Graeco-Latin square.

There is a further step that may be taken toward streamlining generalizations. Until now, the constant sum in each case has been represented by an expression that is an accidental outcome of the matrix design. For instance, in the isomorphic squares of Figures 3 and 5 it is $p + q + r + s$ and $A + B + C + D + a + b + c$, respectively. To standardize this magic constant, all that is needed is to stipulate its value at the outset, say k, and then to complete the generalization so as to comply with this new condition. In the Chernick square of Figure 3, for example, every appearance of s would then be changed to $k - (p + q + r)$. However, if the magic constant is set to zero (or group identity element), k may be eliminated from the matrix and the number of distinct variables reduced by one. This is illustrated in Figure 9, where twin generalizations of *orthogonally*-magic (i.e., magic on rows and columns only) and *diagonally*-magic (i.e., magic on all, including broken, diagonals) squares of order-3 are shown; $k = 0$. All subsequent formulae will appear in this zero-sum form, and of course isomorphism with zero-sum Chernick equivalents is easily demonstrated just as before.

If it seems odd that the matrix can surrender a variable without detriment to its information content, observe that there is indeed a loss. Universality is sacrificed because the reduced matrix now generalizes zero-sum magic squares only. However, the real significance of generalizations lies in their ability to express *intercellular relations*, in which respect nothing has been lost. A nice illustration of this is Lucas's formula of Figure 2, in which a zero-sum square similar to those of Figure 9 is superimposed on a uniform matrix of c 's. The latter contributes nothing to the essential structure, but merely represents the mean value ($k/n = c$) of each cell, where n is the size of the square.

Looked at in this way, an interesting similarity emerges in the construction principle underlying both Lucas's formula (Figure 2) and Bergholt's square. For in both cases, an initial trivial magic matrix becomes *detrivialized* through superposition of magic-preserving zero-sum variable patterns. In Bergholt's case the initial matrix is a Latin square, in Figures 5, 6, 7, and 8 it is a Graeco-Latin square, and in Lucas's formula it is the uniform matrix.

Once this has been glimpsed, it is natural to contemplate the possibility of a uniform matrix-based formula for order-4. My own exploration in this direction has yielded the square seen in Figure 10. Although less economical than other squares, here at last we recover that symmetry in the distribution of variables that intuition tends to anticipate. The elegance of such structures makes it difficult to see them as mere algebraic abstractions; to the receptive eye they become rather, acrostic, palindromic poems written in the language of mathematics. More prosaically, Figure 10 represents a 4×4 counterpart to the 3×3 formula of Figure 2.

Seen from the present perspective, Bergholt's square may now be located roughly halfway along a spectrum of possible formulae, extending from those based on the uniform matrix, to those based on a Graeco-Latin square. Besides these, there are the Chernick-type squares, but

a	$\overline{ab}$	b
$\overline{ac}$	$abcd$	$\underline{bd}$
c	$\overline{cd}$	d

Orthogonally-magic.

$\overline{cd}$	a	$\overline{bd}$
b	$abcd$	c
$\overline{ac}$	d	$\overline{ab}$

Diagonally-magic.

Fig. 9

abg	$\overline{a}df$	$\overline{adf}$	$a\overline{bg}$
$c\overline{b}e$	$\overline{cdg}$	$\overline{c}dg$	$cb\overline{e}$
$\overline{cbe}$	$c\overline{d}g$	$cd\overline{g}$	$\overline{c}be$
$\overline{a}b\overline{g}$	$ad\overline{f}$	$a\overline{d}f$	$\overline{ab}g$

Fig. 10

of course all these different matrices are mathematicallly identical, being only alternative expressions of the same underlying algebriac form. Even so, the construction and classification of such algebraic squares opens as an attractive field of enquiry. Among the more obvious goals of research is a uniform matrix-based version of order-5, while at the other extreme there opens the possibility of 5×5 formula based upon *three* superimposed Latin squares. Incidentally, no 3×3 *diagonal* Latin square exists, but a Chernick square for this order is readily constructed.

The concept of the magic square may of course be extended to matrices of various shapes or dimension, and similarly, the cell entries together with the operation(s) performed upon specific constellations of these in order to yield the magic constant need not be confined to familiar forms. In all such constructions the matrix functions as a purely mnemonic device for indicating a certain set of equations. A study of the algebraic generalizations of such higher forms promises to be rewarding, but, in conclusion, I turn rather in the opposite direction, toward the use of these elementary algebraic techniques in elucidating a peculiar subset of 4×4 squares.

Dudeney's 12 Graphic Types

In the same year that witnessed the appearance of Bergholt's square, H. E. Dudeney published an article including a classification of the 880 'normal' or consecutive-integer magic squares of order 4. This was in *The Queen*, published on January 15th, 1910. The same material was later incorporated into his book *Amusements in Mathematics*. According to this scheme, every normal square belongs to one of twelve "Graphic Types," depending upon the distribution of its eight pairs of so-called complementary numbers, 1 and 16, 2 and 15, . . . , 8 and 9. Figure 11 reproduces the twelve diagrams, where the lines indicate those cell couples occupied by complementary pairs. As a matter of fact, an identical categorization appears in W. S. Andrews' *Magic Squares and Cubes*, which predates Dudeney's article by two years. But although Andrews' has the priority, he blotted his copybook by completely overlooking one of the twelve Types, a slip that obliged him to add a sentence to the 1917 edition of his book acknowledging Dudeney's more thorough analysis.

In a table accompanying his diagrams, Dudeney gave an exhaustive enumeration of all 880 normals under his twelve Types, together with their grouping under the traditional headings of Nasik, semi-Nasik, and Simple; see Table 1. Nasik (or 'pan-diagonal', 'panmagic' or 'diabolic') magic squares are those in which the entries belonging to each of the so-called *broken* diagonals again sum to the magic constant. In a 4×4 square these are the four 'long' broken diagonals *ahkn*, *dejo*, *bglm*, and *cfip*, as well as the two 'short' broken diagonals, *belo* and *chin*; as in Figure 12. Semi-Nasiks are those in which only the latter pair are magic. Simple squares are ordinary magic squares: magic on rows, columns, and two main diagonals, only.

My encounter with Dudeney's work occurred while I was still wrestling with the order-5 square described above, but after that was solved I quickly saw that the Dudeney diagrams created a basis for twelve special algebraic squares. Each Dudeney Type expresses a set of relations that, taken in combination with the usual magic conditions, define a restricted or non-universal

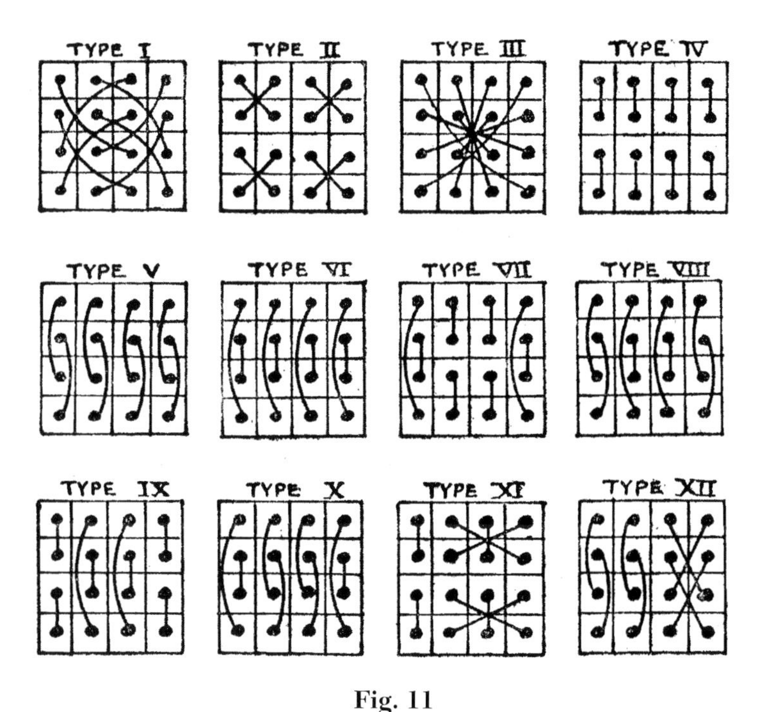

Fig. 11

Type	No. of squares
I (Nasik)	48
II (Semi-Nasik)	48
III "	48
IV "	96
V "	96
VI "	96
VI (Simple)	208
VII "	56
VIII "	56
IX "	56
XI "	8
XII "	8
	880

Table 1

A	*B*	*C*	*D*
E	*F*	*G*	*H*
I	*J*	*K*	*L*
M	*N*	*O*	*P*

Fig. 12

a	b	c	$\overline{abc}$
d	$\overline{abd}$	$a\bar{c}d$	$bc\bar{d}$
$\bar{c}$	abc	$\bar{a}$	$\bar{b}$
$\bar{a}c\bar{d}$	$\overline{bc}d$	$\bar{d}$	abd

Fig. 13 A Chernick-type formula for Type I squares.

generalization. In normal magic squares, the sum of the complementary pairs is half the magic constant, so that when the latter is zero the complements $c1$ and $c2$ are $c1$ and $-c1$ To construct a zero-sum Chernick square corresponding to each Type, we proceed exactly as outlined in the box on page 102, but for every entry a, write $\bar{a}$ in the complementary cell indicated by the Dudeney diagram. Figure 13 shows an example for Type I, which is therefore a generalization covering at least all normal Nasik magic squares of order-4, among others.

It was the work of a single evening to write out all of the twelve generalizations, and exciting to examine the newly disclosed algebraic structures concealed within the Dudeney diagrams for so many years. But at that stage, I had not yet awoken to the advantages of zero-sum notation, with the result that arbitrary magic constants rendered meaningful comparison of different squares impossible. Besides, these matrices suffered from the redundancy and imperspicuity common to the Chernick type, so that revision into reduced form became the obvious next step. This turned out to be a formidable task. The Chernick squares play the part of Ariadne's thread, but trial-and-error guided by experience and intuition remains the only method of advance. Every square is capable of expression in a more or less unlimited variety of forms, so that each underwent several stages of refinement before reaching completion. Finally, in order to maximize comparability of squares, careful consideration had to be given to the allocation of letter symbols. In Figure 14 can be seen the "fruit of pensive nights and laborious days." Here, it will be found, lie answers to some of the intriguing questions concerning those 880 squares first brought to light by the labours of Frénicle de Bessy in 1693.

Reading the Runes

In the same way that certain properties common to every 4×4 magic square can be read off from a universal formula, so it is with the twelve Dudeney Types. Here I confine myself to three main observations.

1 As already noted, not only are Type-I squares Nasik (as Dudeney was well aware), the sum of the entries

Fig. 14

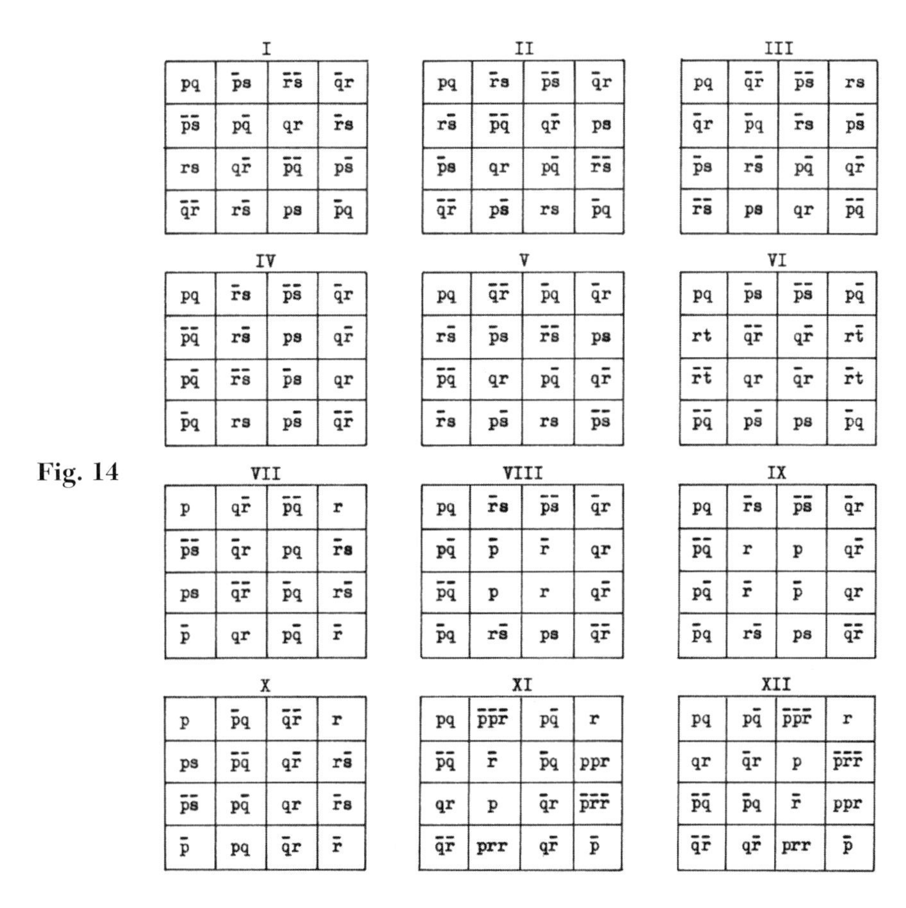

in any 2×2 block of cells is again equal to the magic constant, or zero in this case. This applies equally when the square is viewed as *toroidally-connected*, which is to say, when the top and bottom rows, as well as the left- and right-hand columns are treated as adjacent to each other. Counting up, we find 16 of these zero-totalling 2×2 blocks, corresponding to the letters in Figure 12 as follows: *abef, bcfg, cdgh, adeh, efij, fgjk, ghkl, ehil, ijmn, jkno, klop, ilmp, abmn, bcno, cdop, admp.* Magic squares that exhibit this property are said to be *compact*. Hence, among other things, the message contained in Figure 14 (I) is that every Type-I square (i.e., every normal square whose complementary pairs are distributed as shown) is both Nasik and compact. However, we can go somewhat further than this.

In the same way that the twelve squares in Figure 14 are non-universal formulae based upon Dudeney's Graphic Types, so can Chernick generalizations describing the structure of other special kinds of square be constructed. For example, such a formula can be made to describe Nasik squares only, which is to say, Nasik squares that are not necessarily *normal*. As before, the procedure followed is the same as that shown in the box on page 4, except that, in addition to the usual magic properties, now every broken diagonal is assumed to be magic also. The result is identical to Figure 13. This is of interest because, as the reader can check, Figure 13 turns out to be isomorphic with the Type-I formula of Figure 14. But this shows us that *every* 4×4 Nasik square, normal or otherwise, is always comprised of 8 complementary pairs of entries, and is always compact, a finding that can alternatively be validated by constructing a Chernick generalization of all 4×4 compact magic squares. The result is again identical to Figure 13, which again shows that the three characterizations, Type I, Nasik, and compact, define exactly the same entity.

2 It will not have escaped readers that the set of 16 algebraic terms appearing in the twelve formulae are in several cases the same. As already pointed out, "careful consideration had to be given to the allocation of letter symbols." Specifically, the twelve squares fall into four classes sharing identical cell entries:

Class 1 : Types I, II, III, IV, V, (and that subset of Type-VI when $t = s$ or $-s$)
Class 2 : Type VI (excluding the above subset members)
Class 3 : Types VII, VIII, IX, X
Class 4 : Types XI, XII

Clearly, the entries of a magic square belonging to a given Class may always be transposed to form a new square of different Type in the same Class. Hence, the entries in a Type-XI square, say, can always be rearranged to yield a Type-XII square, and vice versa, while the entries in a Type-V square, for example, can always be rearranged so as to form a Nasik, or Type-I square, and vice versa. This is no trivial finding, since it draws together and explains the entire family of all such magic rearrangements, as has never before been possible. Moreover, it obviates the necessity of having to prove any particular case, such as that found in [6], for example, in which the author devotes two pages to proving that Type-I squares are always transposable into Type III, and vice versa. Figure 14 subsumes all such demonstrations, and more besides, in a single diagram.

Closer examination of the 16 algebraic terms comprising Class 1 reveal them as the entries of an addition table, such as that in Figure 15 left. This is of interest because the entries in Lucas's 3×3 formula of Figure 2 also belong to an addition table (Figure 15 right), a fact that finds explanation in the Parallelogram Theorem that is introduced in *The Lost Theorem*, here reproduced in Appendix V.

Less obvious is that the entries of Types XI and XII are really that special case of Class 1 when $s = pr$. Class 4 magic squares are thus always transposable into Class 1 Types, so that the entries in a Type-XI or XII square can also be rearranged to yield a Type-I, or Nasik square, say.

3 Dudeney's table in Figure 11 shows that some normal squares of Type-VI are semi-Nasik (96) and some Simple (208). But if now in Figure 14 (VI) we look at the special cases when $t = s$, or $t = -s$, respectively, the resulting matrix becomes as in Figure 16.

The 16 entries occurrring in these Type-VI algebraic

$+$	p	$\bar{p}$	r	$\bar{r}$
q	pq	$\bar{p}q$	qr	$q\bar{r}$
$\bar{q}$	$p\bar{q}$	$\overline{pq}$	$\bar{q}r$	$\overline{qr}$
s	ps	$\bar{p}s$	rs	$\bar{r}s$
$\bar{s}$	$p\bar{s}$	$\overline{ps}$	$r\bar{s}$	$\overline{rs}$

$+$	a	0	$\bar{a}$
bc	abc	bc	$\bar{a}bc$
c	ac	c	$\bar{a}c$
$\bar{b}c$	$a\bar{b}c$	$\bar{b}c$	$\overline{ab}c$

Fig. 15

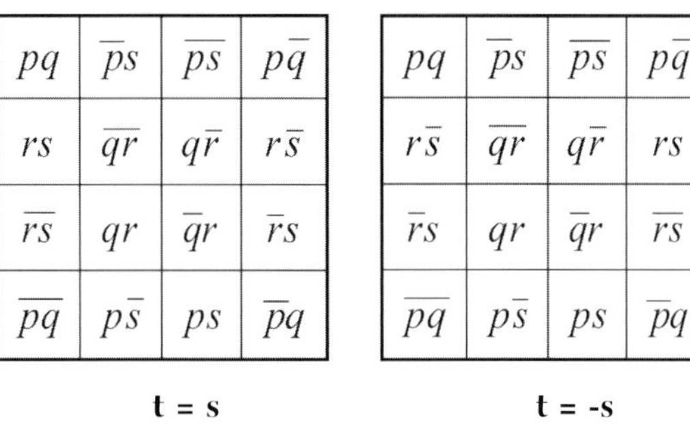

Fig. 16

squares are now identical to those that appear in the formulae for Types I – V. Moreover, the square on the right is semi-Nasik, while the square on the left is what I call "skewed semi-Nasik," a new category I introduce here that will prove helpful. Using Chernick equivalents, it is not difficult to prove that these squares are indeed general formulae for Type-VI semi-Nasiks and skewed semi-Nasiks, respectively. Every one of Dudeney's 96 normal semi-Nasiks must, therefore, be an instance of the right-hand formula in Figure 16.

Figure 17 indicates the two extra zero-totalling sets of four entries that characterize these squares.

	X	Y	
Y			X
X			Y
	Y	X	

skewed semi-Nasik

	X	Y	
X			Y
Y			X
	Y	X	

semi-Nasik

$4X = 4Y$ = magic constant

Fig. 17

A semi-Nasik square is one in which the 4 numbers in each of the two short broken diagonals sum to the magic constant. We define a skewed semi-Nasik as a Type VI magic square in which the 4 cells marked X, and the 4 marked Y, again sum to the magic constant.

A computer program that checked to see how many distinct magic squares can be formed using the 16 algebraic terms of Class 1, finds 528, rotations and reflections not counted. But this is 96 greater than the sum of Dudeney's totals for Types I–V, plus his 96 Type VI semi-Nasiks: 48(I) + 48(II) + 48(III) + 96(IV) + 96(V) + 96(VI semi-Nasik) = 432. This reveals that Dudeney's 208 'Simple' normals must include 96 skewed semi-Nasiks, a deduction that fits nicely with the fact that every Type-VI semi-Nasik can be rearranged to yield a Type-VI skewed semi-Nasik square.

Conclusion

The year of writing marks the fiftieth anniversary of Henry Ernest Dudeney's death, exactly seventy years following the joint appearance of his and Ernest Bergholt's work discussed above. Whether they had any inkling of the eventual importance of being Ernest is unknown, but Dudeney's earnestness in describing Bergholt's work as "of the greatest importance" has already been noted. In any case, the combination of their techniques as exemplified in the squares of Figure 14 appropriately commemorates their oft neglected contribution to the theory of 4×4 magic squares.

Every algebraic square, universal or otherwise, corresponds to an x-ray photograph exposing the skeletal structure underlying, yet concealed, by the arithmetical exterior of numerical magic squares. The differing approaches of Bergholt and Chernick toward identifying this structure afford perfect illustrations of the two methods of science as depicted in the epistemology of the contemporary philosopher Karl Popper [7]. On the one hand, Chernick comes up with an algorithm by means of which the sub-surface contours may be *deduced*. On the other, Bergholt relies upon intuition aided by experience in leaping to a bold *conjecture* as to the hidden form. His mistake lay in believing his result verified, when, in fact, at that stage, it had not yet superceded the status of a falsifiable hypothesis; see Figure 18. My own contribution to this muddle has been to show how the logic of the former may be used to acquit the guesswork of the latter.

My essay will have served its purpose if it succeeds in drawing the reader's attention to a hitherto neglected area in the theory of magic squares. The matter is recondite, curious and exciting, rich in problems and possibilities, and very largely unexplored. In the past, although more fundamental, algebraic generalizations have been almost totally eclipsed by their arithmetic instances. This is regrettable, for some of these matrices possess a prismatic quality whose transparency uniquely

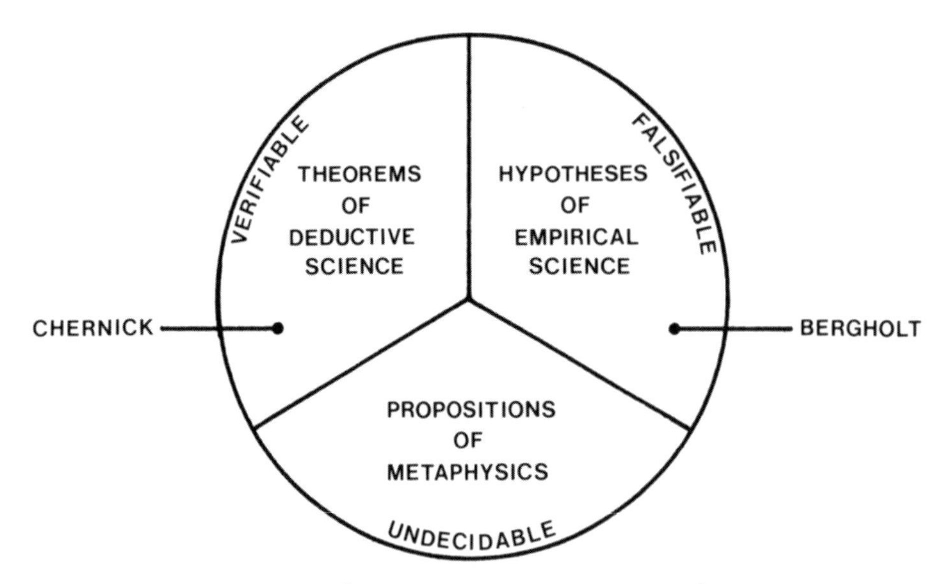

Fig. 18 Karl Popper's tri-partite epistemology

illumines and magnifies the crystalline structure of magic squares.

References

[1] *Nature*, No 2117, Vol 83, pp 368-9

[2] pub. by Nelson, London, 1917, reprinted by Dover Books, 1958.

[3] Twelfth edition pub. by U. of Toronto Press, 1974.

[4] *Am. Math. Monthly*, 1938, Vol 45, pp 172-5.

[5] *Récréations Mathématiques*, Paris, 1894, Vol IV, p. 225.

[6] Johnson, Jr., Allan Wm. *Journal of Recreational Mathematics*, Vol 12, No 3, pp 207-9.

[7] Popper, Karl, *The Logic of Scientific Discovery*, Routledge.

Appendix III
New Advances with 4x4 Magic Squares

Introductory Note

The article to follow, *New Advances with 4×4 Magic Squares* presents the results of two strands of research carried out at various periods during the 1990's. This work has never before appeared in print, nor indeed ever been submitted for publication, although in recent years Harvey Heinz was kind enough to make it available on his excellent website: http:// www.magic-squares.net. To the best of my knowledge, the matter of *fertility* here examined has never previously been explored. It would be nice to think that some of the intriguing questions thrown up by the findings here recorded will stimulate others to pursue these matters more deeply. In my own view, the most obvious goal of future research would be to discover a non-brute-force algorithm whereby the fertility of a set of 16 randomly chosen integers can be determined immediately, without need of a computer.

New Advances with 4x4 Magic Squares

Introduction

One of the best known results in the magic square canon is Bernard Frénicle de Bessy's enumeration of the 880 'normal' 4×4 squares that can be formed using the arithmetic progression 1, 2 , . . . , 16. A natural question this suggests concerns non-normal squares: Is 880 the largest total attainable if *any* 16 distinct numbers are allowed?

The answer is no. A computer program that will generate every square constructable from any given set of integers has identified 1,040 distinct squares using the almost arithmetic progression: 1, 2, 3, 4, 5, 6, 7, 8, 10, 11, 12, 13, 14, 15, 16, 17. Note the doubled step from 8 to 10. The constant row-sum now becomes 36. Extensive trials with alternative sets make it virtually certain that 1,040 is the maximum attainable (or 8 × 1,040 = 8,320, if rotations and reflections are included), although an analytic proof of this assumption is lacking. A listing of the 1,040 squares can be had on request via email at lee.sal@inter.nl.net.

But if 1,040 is the maximum, what of lower totals? Can a set of numbers be found that will yield any desired total under 1,040? If not, what totals are possible? In the absence of mathematical insight, trial-and-error was the only way to find out. The results of computer trials on thousands of different sets of small integers are presented below. Again, although proof is lacking, there exist empirical grounds for supposing that every possible total has been identified. Conceivably, one or two may have escaped the net. In any case, the wealth of new data represented by these findings ought to provide a starting point for further researches in the field for years to come. Some preliminary comments will be helpful in understanding these results.

Fertility and Set Structure

We define the *fertility*, *f*, of a set of 16 distinct elements, usually integers, as the total number of 4×4 magic squares it can produce, rotations and reflections included, *divided by* 32. This is because there are magic-preserving rearrangements besides rotations and reflections that are applicable to any square such that it always has 31 variants that we may count as equivalent. For example, the two

T1

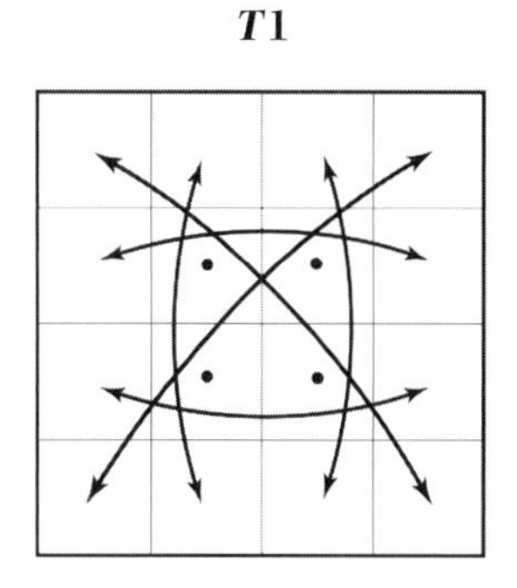

T2

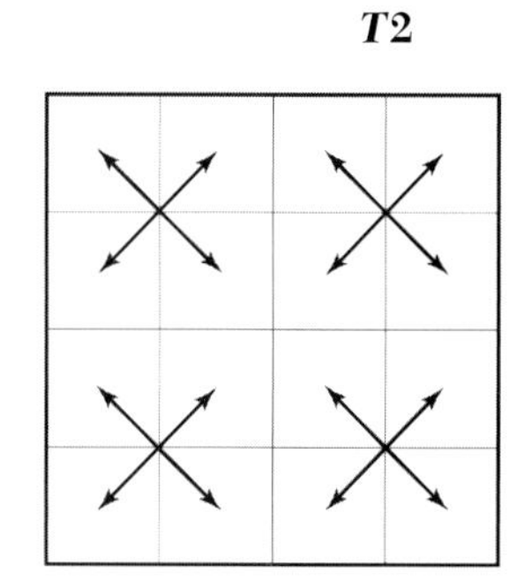

Fig. 1

transformations shown in Figure 1 can be combined with rotations and reflections to produce all 32 squares.

Thus, for the above-mentioned set, f = 8,320÷32 = 260, the putative maximum, while for a set of 16 consecutive integers, f = 880 × 8 ÷ 32 = 220. Besides these two cases, integer sets exhibiting around one hundred different fertility values have been found, as listed in Tables 1 and 3.

Curiously, among these new finds is a second set yielding f = 220, but not using consecutive integers. In other words, for every one of the 880 normal squares, there exists a unique counterpart square using 16 numbers that do not form an arithmetic progression. An example is shown in Figure 2. The constant row sum is now 38.

2	16	12	8
18	6	4	10
7	15	13	3
11	1	9	17

Fig. 2

It is interesting to note that whereas 4 distinct numbers adding to 34 can be chosen from {1, 2, . . . , 16} in 86 different ways, there are but 76 ways to choose 4 numbers adding to 38 from the above set. Still more surprising is the outcome for the set yielding 1,040 squares: 4 numbers adding to 36 can be chosen in just 80 ways. The relation between the number of available magic 4-partitions and the number of squares produced would thus seem to be a lot looser than might have been supposed.

In comparing different sets it is useful to look at their *structure* rather than the numbers occurring. This is because the fertility of a set is unaffected by adding a constant to each member, or by multiplying each member by a constant, so that different numbers need not imply different fertilities. Sets are thus better described by the sequence of 15 *differences reduced to lowest terms* occurring between adjacent members when ordered by magnitude. And since differences in the sets considered here are always less than ten, a single digit suffices for each. Thus, the two sets:

$$\{-3, -1, 1, 3, 5, 7, 9, 13, 15, 17, 19, 21, 23, 25, 27\}$$

and $$\{1, 2, 3, 4, 5, 6, 7, 8, 10, 11, 12, 13, 14, 15, 16, 17\}$$

are both counted as particular instances of the structure: 111111121111111, which is our canonical representation of all sets for which $f = 260$. The two set structures yielding $f = 220$ are then 111111111111111, representing arithmetic series, and 111211111112111, as exampled in the magic square of Figure 2.

The foregoing were all examples of palindromic or *symmetric* sets, which are those composed of 8 conjugate pairs of numbers having equal sums, such as 1 + 16, 2 + 15, . . . 8 + 9 in normal squares. If we think of the numbers as points along the real number line, then conjugate pairs are points equidistant about a common centre, which is half their sum. *Asymmetric* sets are non-palindromic, but since the fertility of a set is unchanged by multiplying each member by –1, the fertility of an asymmetric set is the same as that of a set of same structure but reversed in order.

These remarks are sufficient to explain Table 1, which lists 66 asymmetric sets yielding fertility values in the range 1 to 71. No asymmetric set with a fertility outside this range was discovered; the five gaps indicate cases for which no asymmetric sets were found. Asterisks indicate fertility values for which symmetric sets also exist. Table 1 extends beyond $f = 71$ to include the asterisk at $f = 76$, beyond which no further symmetric set is found until $f = 132$, as seen in Table 3. These are among many findings that have as yet no theoretical explanation.

The set listed in each case is usually the *smallest*, in the sense that the sum of its differences is least; but not always. For instance, 111112111212111 also yields $f = 1$, the sum of its differences being the same as the set structure listed for $f = 1$ in Table 1. In general, the lower the value, the more asymmetric sets with the same fertility exist. For instance, among all possible sets with structures in the range 111111111111111 to 222222222222222, we find 455 yielding $f = 1$, the totals for succeeding f-values thereafter diminishing as shown in Table 2.

Table 1: Fertility values for asymmetric sets

f	set structure	f	set structure
1	111111211112211	45	111211111113111
2	111111212112111	46	111111121114111
3	111211111311111	47	111111111114111
4*	111111111131121	48*	111111111312131
5	111111111112131	49	111211111211121
6	111111111121211	50	111111111111151
7	111111111211121	51	121111111112113
8*	111111111212111	52*	111113131111131
9	111111112122111	53	111111111112111
10	111111211121121	54	111111121112111
11	111111111111321	55	311111212211122
12*	111111111121122	56*	111121211111212
13	111121111111112	57	111111111115111
14	111111211111211	58	121111111111113
15	111111111122211	59	111111122111112
16*	111111111111212	60*	111111141113111
17	111111123111121	61	311111111123212
18*	111112111112121	62	111111112211122
19	111111111111131	63	111111111113111
20*	111111111221221	64*	111111142111112
21	111112111311121	65	none
22*	111111111113115	66	none
23	111111111111115	67	121111111112133
24*	111112222121112	68*	
25	111112111111311	69	111211111114111
26*	111113111211311	70	none
27	211111111111322	71	111111112111112
28*	111111111112511	72*	
29	111112111111222	73	none
30*	111111121111321	74	none
31	311111111111222	75	none
32*	111111321112111	76*	
33	none	>76	none
34*	112111111111122		
35	none		
36*	311111111211321		
37	121111112111134		
38*	111212121112121		
39	121121241211212		
40*	111311111114111		
41	111211111111123		
42*	111112131111121		
43	111111111112212		
44*	111112111111121		

* = symmetric set exists

Table 1

f	1	2	3	4	5	6	7	8	9	10	11	12	13	14	15
	455	494	248	236	117	118	83	95	53	27	19	53	6	15	13
f	16	17	18	19	20	21	22	23	24	25	26	27	28	29	30
	19	3	5	3	19	1	2	3	21	26	1	0	15	1	2

Table 2

A	$B+a$	$C+b$	$D+c$
$C+c+x$	$D+b$	$A+a$	$B-x$
$D+a-x$	C	$B+c$	$A+b+x$
$B+b$	$A+c$	D	$C+a$

Fig. 3

This is not difficult to understand. Figure 3 shows a general formula that describes the structure of every 4×4 magic square:

The number of magic squares that can be formed using the 16 algebraic terms in the formula is clearly the same as the number of magic-preserving rearrangements applicable to *any* numerical magic square, and no more: 32. The fertility of this algebraic set is thus 1, but sets of greater fertility must represent particular instances of the above, satisfying still more stringent conditions, and to that extent will be scarcer. Hence the tapering off in totals as fertilities increase.

Turning now to Table 3, we find a total of 63 symmetric set structures, all of them yielding even fertility values in the range 4 to 260. No set with an odd-valued fertility was found. Why is it that the fertility of symmetric sets is always even?

A tempting answer might seem to lie in the fact that symmetric squares come in complementary pairs, such as the two shown in Figure 4.

1	8	13	14
15	12	7	2
16	11	6	3
4	5	10	17

17	10	5	4
3	12	11	16
2	7	6	15
14	13	8	1

Fig. 4

Switch the conjugate number pairs in one square to get the other, its so-called *complement*. The trouble is that in this example the complement of each square is the same thing as its rotation by 180 degrees. But this means that if one square is among those in a set of f distinct specimens, then its complement (i.e., its rotation) will *not* be, so that this attempt to answer the original question fails. However, a proof that the fertility of symmetric sets is always even is derivable, although too space-consuming to be included here.

Returning to Table 3, note that sets having fertilities in the range 1 to 76 have been found, with about 10 gaps, those appearing in the right hand column all being multiples of 4. Beyond these there is a wide gap up to 132, followed by most even values from 132 to 220, again with about 10 gaps. No sets with $220 < f < 260$ have been found. Both sets for $f = 220$ are recorded in the list.

The criteria used in selecting the particular set structure shown for each fertility value in the Tables differs from case to case. Frequently, there exist many alternative structures with the same fertility that could have been shown instead. Many of the results are due to Saleem al-Ashhab, a Jordanian email correspondent. Saleem was excited by my 1,040 squares find, and soon had his PC running night and day in search of sets with new f values. In such cases I received only his results, but without details of the search procedure used. Still other results are due to programs of my own, one of which checked blocks of sets in systematic order, counting the number of squares produced by each in turn and recording any new values found. I might mention that, thanks to insights due to Saleem, the fertility

Table 3: Fertility values for symmetric sets.

f	set structure	f	set structure
2	none	4	11111223
6	none	8	11212121
10	none	12	11121221
14	none	16	11112221
18	22443111	20	11111213
22	11111331	24	22121211
26	11111131	28	11122211
30	11222111	32	11111121
34	22121111	36	11112111
38	11222112	40	21121111
42	22121131	44	13111231
46	12321111	48	41121122
50	none	52	12111132
54	none	56	11111221
58	none	60	13111212
62	none	64	11111212
66	none	68	12111112
70	none	72	21112112
74	none	76	11211122
78-130	none	132	23111327
134-142	none	144	23111325
146	23111323	148	11131117
150-4	none	156	12161211
158	12141212	160	13111315
162	21111124	164	11131113
166	none	168	31121132
170	12141211	172	22111224
174	21111123	176	42112111
178	11111115	180	13111312
182	22212111	184	21212122
186	11141111	188	11131111
190	31211111	192	23211112
194	11121113	196	21111121
198	11111113	200	13121111
200	12111211	202	13111111
204	22111111	206	11111114
208	11121112	210	none
212	32111111	218	21111111
220	11111111	220	11121111
260	11111112	>260	none

First half only of palindromic set structures are shown; i.e. 11111112 is short for 11111121111111.

Table 3

of a chosen set of 16 integers took the final version of our computer program less than 2 seconds to compute on a Pentium II machine. In another approach, sets composed of 16 random integers were generated one after the other, their fertility determined and the result stored. Haphazard as this may seem, following an initial flood, with the passage of time the number of fresh fertility values found slowed to a dribble and then finally dried up completely, even following whole days of running time. Hence my confidence in the probable completeness of the results here presented. It is worth mentioning that trials on sets using non-integral numbers produced no fertility values different to those recorded here.

Non-Normal Graphic Types

A further familiar landmark in the lore of 4×4 squares is H. E. Dudeney's classification of the 880 normals into 12 "graphic types," depending upon the different possible positions of the eight conjugate (or complementary) pairs, 1–16, 2–15, . . . , 8–9 [*Amusements in Mathematics*, p. 120]. The twelve Types are seen in Figure 5.

W. S. Andrews gives the same classification in the 1908 edition of his *Magic Squares and Cubes*, p.180, but overlooked one type, admitting in later editions that Dudeney had supplied the twelfth. I suspect Andrews may have felt disgruntled by this because he went on to say that Dudeney's 12 types "probably cover all types of 4×4 magic squares." There is no "probably" about it. Dudeney's exhaustive enumeration of every possible type is detailed in *The Queen* for January 1910.

A second natural question prompted by the above concerns non-normal squares: Do Dudeney's 12 types account for every possible pattern if squares using *any* 8 conjugate pairs (i.e., pairs of equal sum) are allowed?

The answer has been known in Japan for more than 40 years. If the question were ever posed, I can find no trace of it in the magic square literature of the West.

The first non-Dudeney type square was discovered by Gakuho Abe in 1957. Abe was investigating a special kind of 7×7 square containing embedded 3×3 and 4×4 magic squares in opposite corners. An example is shown in Figure 6.

The 4×4 squares were thus constructed using numbers in the range 1 to 49, and although non-normal, could yet include 8 conjugate pairs (1, 49; 2, 48; 3, 47; 4, 46; 6, 44; 7, 43; 8, 42; 9, 41, here). Abe noticed that in a few cases their distribution was non-Dudeney, as above. He went on to

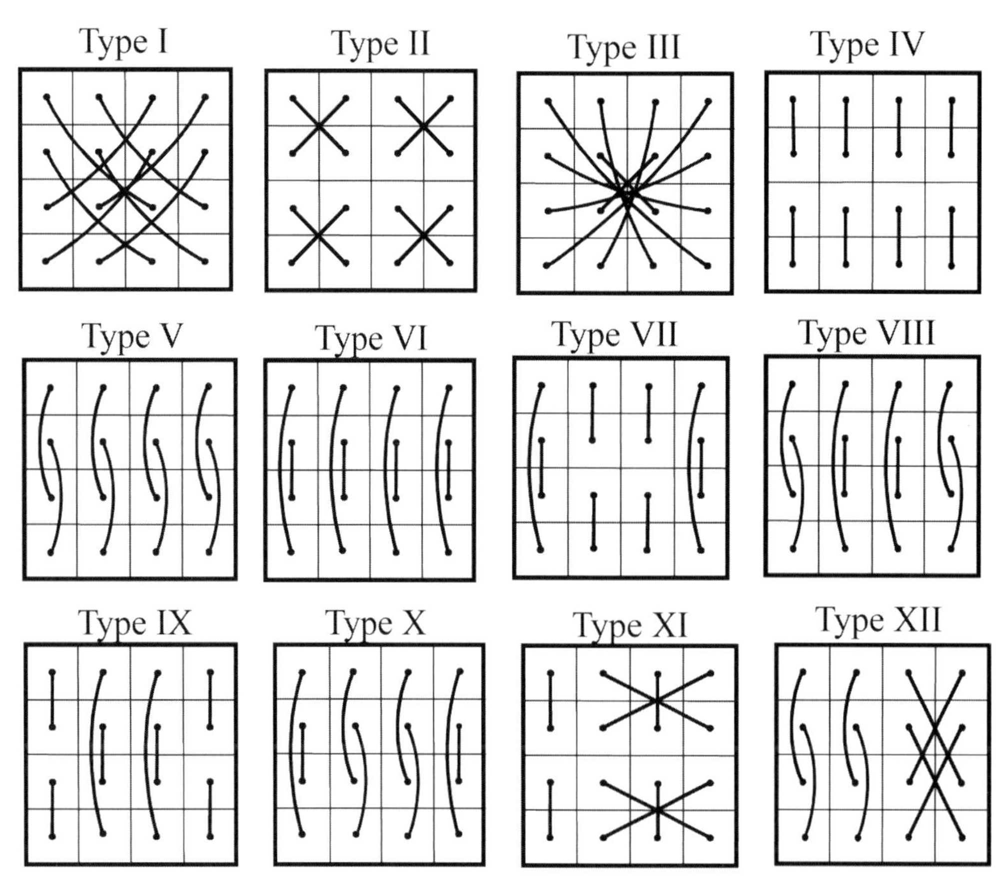

Fig. 5

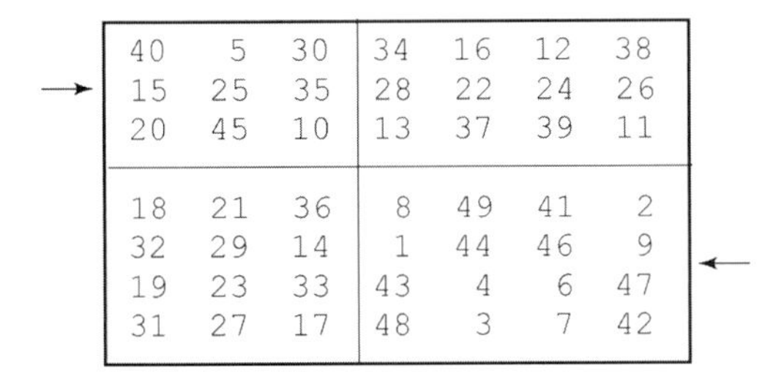

40	5	30	34	16	12	38
15	25	35	28	22	24	26
20	45	10	13	37	39	11
18	21	36	8	49	41	2
32	29	14	1	44	46	9
19	23	33	43	4	6	47
31	27	17	48	3	7	42

Fig. 6

discover six non-Dudeney types that later appeared in his article "Yon Ho jin no taiikei no zensaku," or "Complete Complementary Pair Models of Order Four Magic Squares."

Nineteen years later an interesting response appeared. In 1976 Tomiya Yokose published a systematic study of all possible graphic types in which he identified 30 in all, Dudeney's 12 included [Sugei no pazuru (= "Mathematical Puzzles"), No.92, Sept-Oct., Showa 51, pp 17-25]. In my humble opinion Yokose's work is a milestone in the development of the theory of 4×4 squares. As before, however, news of it seems never to have reached the West.

I am in debt to Nyr Indictor of New York State, and Guido de Mey of Brussels for translations of Abe's and Yokose's articles.

Still later I looked afresh at this question and, with a lot of help from Michael Schweitzer, discovered four types that Yokose had missed. A computer search was employed; it would be tedious to describe the method here. Following on from Dudeney's 12, I numbered the new types 13 to 34, as seen in Figure 7.

The ordering of Types 13 – 34 is partially random, partially systematic. It can be shown that whatever

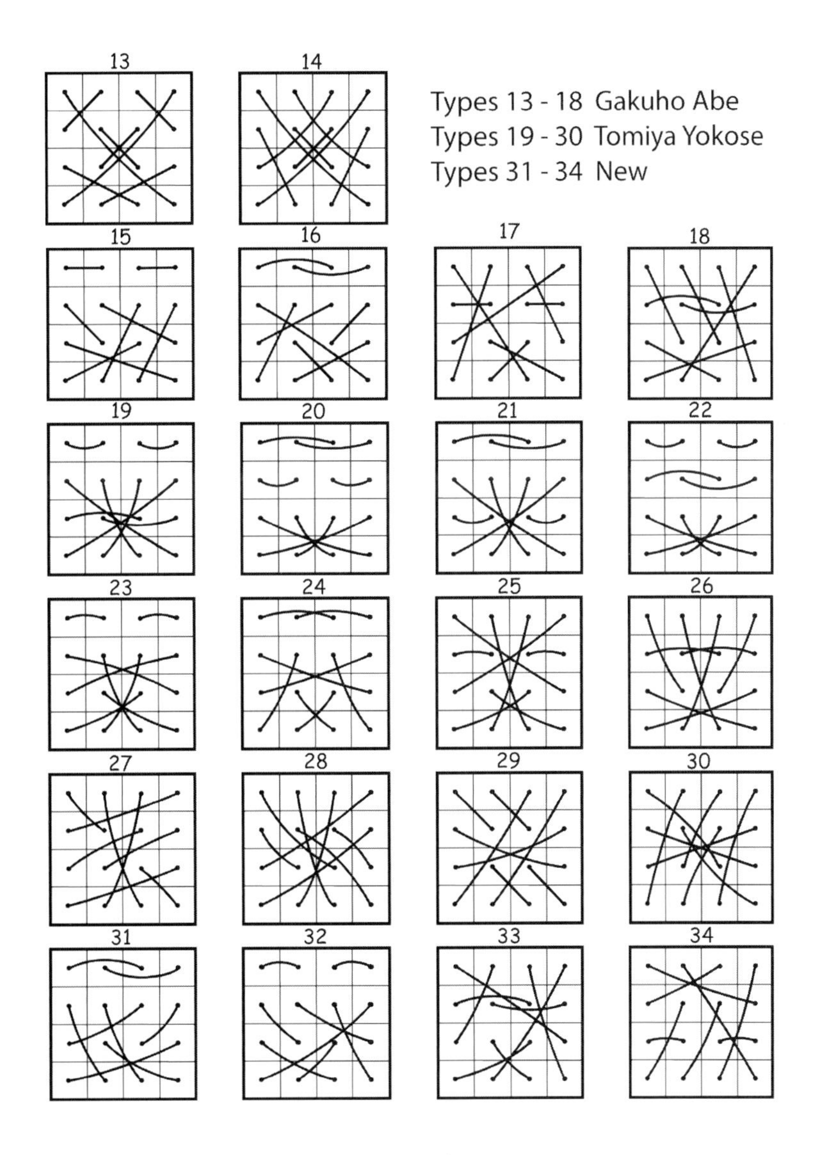

Fig. 7

numbers appear in a Type-13 square, they can always be rearranged so as to yield a Type-14 square, and vice versa. The same is true for any four Types appearing in the same horizontal row in the above diagram. Specifically, Types occupying the 2nd, 3rd and 4th columns can be derived from that in the 1st column by means of the transforms $T1$, $T2$, and $T1 \times T2$, respectively, shown at the beginning of this article.

A general formula for magic squares of order 4 has already been shown. Each of the above Types represents a subset of magic squares whose properties can be described by a more restricted formula. Since Types in the same row above are merely rearrangements of each other, a single formula for one square taken from each row suffices to cover all cases, as provided in Figure 8.

The magic constant in the formulae is zero. Add k to every cell to in other to generalize squares having any magic constant, $4k$.

Formulae describing the structure of Dudeney's original twelve Types can be found in my 1980 article "*Magic Formulae.*"

Finally, it may be of interest to know that examination of the 1,040 squares discussed above shows them all to belong to Types 1 through 12, an example of each being as shown in Figure 9.

Type 13

$\bar{a}c$	$\overline{abc}$	$ab\bar{c}$	ac
abc	$b\bar{c}$	$\overline{bc}$	$\overline{ab}c$
$a\overline{bc}$	bc	$\bar{b}c$	$\bar{a}b\bar{c}$
$\overline{ac}$	$a\bar{b}c$	$\bar{a}bc$	$a\bar{c}$

Type 23

$\overline{5a+b}$	$5a+b$	$\overline{5a-b}$	$5a-b$
$4a-b$	$\overline{a-b}$	$a+b$	$\overline{4a+b}$
$4a+b$	$\overline{3a+b}$	$3a-b$	$\overline{4a-b}$
$\overline{3a-b}$	$\overline{a+b}$	$a-b$	$3a+b$

Type 15

$7a$	$\overline{7a}$	$21a$	$\overline{21a}$
$\overline{11a}$	$\overline{3a}$	a	$13a$
$\overline{5a}$	$11a$	$\overline{9a}$	$3a$
$9a$	$\bar{a}$	$\overline{13a}$	$5a$

Type 27

$4a-b$	a	$\overline{7a}$	$2a+b$
$\overline{2a+b}$	$\overline{4a-b}$	$2a-b$	$4a+b$
$\overline{2a-b}$	$\overline{4a+b}$	$6a+b$	$\bar{b}$
b	$7a$	$\bar{a}$	$\overline{6a+b}$

Type 19

$4a-b$	$\overline{3a+b}$	$2a+b$	$\overline{3a-b}$
b	$\overline{5a-b}$	$2a-b$	$3a-b$
$\overline{2a-b}$	$3a+b$	$\bar{b}$	$\overline{a+b}$
$\overline{2a+b}$	$5a-b$	$\overline{4a-b}$	$a+b$

Type 31

$7a$	$\overline{3a}$	$\overline{9a}$	$5a$
$21a$	$\overline{11a}$	$3a$	$\overline{13a}$
$\overline{7a}$	$13a$	$\overline{5a}$	$\bar{a}$
$\overline{21a}$	a	$11a$	$9a$

Fig. 8

I

-8	-1	2	7
4	5	-6	-3
-2	-7	8	1
6	3	-4	-5

II

-8	-5	6	7
5	8	-7	-6
-1	-4	3	2
4	1	-2	-3

III

-8	-1	4	5
6	3	-2	-7
7	2	-3	-6
-5	-4	1	8

IV

-8	-5	6	7
8	5	-6	-7
-2	-3	4	1
2	3	-4	-1

V

-8	-5	8	5
6	7	-6	-7
-2	-3	2	3
4	1	-4	-1

VI

-8	-7	7	8
2	6	-6	-2
5	-3	3	-5
1	4	-4	-1

VII

-7	-5	8	4
1	5	-8	2
-1	-3	6	-2
7	3	-6	-4

VIII

-8	-6	8	6
-1	7	-7	1
5	-3	3	-5
4	2	-4	-2

IX

-8	-5	7	6
8	1	-3	-6
2	-1	3	-4
-2	5	-7	4

X

-7	-5	5	7
6	8	-6	-8
-2	-4	2	4
3	1	-1	-3

XI

-8	6	-2	4
8	-4	2	-6
1	-5	7	-3
-1	3	-7	5

XII

-8	-6	8	6
2	4	-2	-4
7	-3	1	-5
-1	5	-7	3

Fig. 9

Appendix IV
The Dual of the *Lo shu*

Introductory Note

A couple of blows have already been aimed at the iconic reputation of the *Lo shu* magic square in the course of Concluding Remarks. In The Dual of the *Lo shu* which started life as the first part of a longer, unfinished article on ambimagic squares, I seek to deliver the coup de grâce. The *Lo shu*, it emerges, is not the unicorn we have always taken it to be, but merely one half of a pantomime horse. Ah, but is it the back half or the front? In any case, it will be no use locking the stable door after the horse has bolted it. Because a horse that bolts doors can only *be* a pantomime horse, which is what I set out to prove. The essay, not previously published, was written in 1999.

The Dual of the *Lo shu*

"Various .. authors .. have asserted that the earliest magic square known [outside China] *was given by Theon of Smyrna, a Neo-Pythagorean, about 130 A.D. However, if they had looked up this so-called 'magic square' of Theon's, they would have seen that it was not a magic square in any sense of the term."*
Schuyler Cammann [1]

"This array of numbers [Theon's square] *has therefore not the slightest thing to do with magic squares and deserves no place in the history of magic squares."*
W. Ahrens [2] [author's translation]

Duality Denied

Cammann and Ahrens, the two authors quoted above, ought to know what they are talking about. The first has written several oft-cited monographs on ancient Chinese magic squares, while the second is the author of a celebrated German two volume classic of recreational mathematics, *Mathematische Unterhaltungen Und Spiele.* So what exactly is this thing, 'Theon's square', against which they are united in their contempt?

Imagine a text on Taoist philosophy that expounds the Yin principle but without mentioning the Yang. Or an abridgment of Lewis Caroll's *Through the Looking Glass* that introduces Tweedledee yet omits any reference to Tweedledum. An unlikely scenario perhaps, and yet something equivalent to it is exactly the state of affairs in the realm of 3×3 magic squares.

"The literature on magic squares is vast," wrote Martin Gardner. An exaggeration perhaps, but as I have said myself in The Lost Theorem, innumerable books and articles on magic squares begin with a discussion of 3×3 types, the properties of which have long been regarded as completely understood [3]. And in almost all of those books and articles will be found the oldest and most famous magic square of all. It is the 3×3 square of Chinese origin known as the *Lo shu*, a prototype so revered among aficionados that it enjoys the status of an iconic emblem; see Figure 1. The weird thing about all these writings on the *Lo shu*, however, lies in what they don't say. I refer to the total absence of any mention of the *Lo shu's* twin or duplicate square, or as we should say in mathematical language, its *dual*.

8	1	6
3	5	7
4	9	2

Fig. 1 The *Lo shu*.

I can imagine that this reference to the dual of the *Lo shu*, innocuous enough to the lay reader, will be greeted with incredulity on the part of more seasoned students, none of whom will have heard of it for exactly the reason just outlined. That is, to the best of my knowledge, there exists not a *single* reference to the existence of this dual anywhere in the 'vast' literature on magic squares. On the contrary, there are even experts on the subject who will deny the contingency outright. I have already quoted two of them. They are our two learned savants, Cammann and Ahrens, the authors for whom Theon's square is such a pain in the neck. So what (to return to our starting point) *is* this object, *Theon's square*?

Theon's Square

The definition of a magic square means that it occupies a halfway house between *orthomagic* squares and *panmagic* squares. In the former, sometimes called *simple*, or *semimagic* squares, the diagonal requirement is dropped, with the the rows and columns, or *orthogonals* alone needing to show equal sums. In the latter, also known as *panmagic*, *pandiagonal* or *Nasik* squares, the totals in *every* diagonal, including the so-called 'broken' diagonals, are the same as those in all of the orthogonals. In Figure 2, the broken diagonals are occupied by *afh*, *cdh*, *ibd*, and *gbf*.

a	*b*	*c*
d	*e*	*f*
g	*h*	*i*

Fig. 2

In fact, although panmagic squares exist for all higher orders, a 3×3 panmagic square using nine distinct numbers is impossible to construct.

Broken diagonals can however be mended by bringing together the top and bottom edges of the square to make a cylinder which is then stretched and bent smoothly in a circle until its ends meet to form a closed tube or torus. In a square array of 3×3 thus 'toroidally connected,' the left and right hand edges coincide, as do those at top and bottom, with the result that every diagonal, like every orthogonal, becomes a closed loop of 3 cells. The eight immediate neighbors of each cell are then the same as if the square were surrounded with copies of itself, as in using it as a tile to cover the plane. The closed loops are then reflected in repeated cycles of 3 elements along horizontal, vertical, and inclined paths as seen in Figure 3.

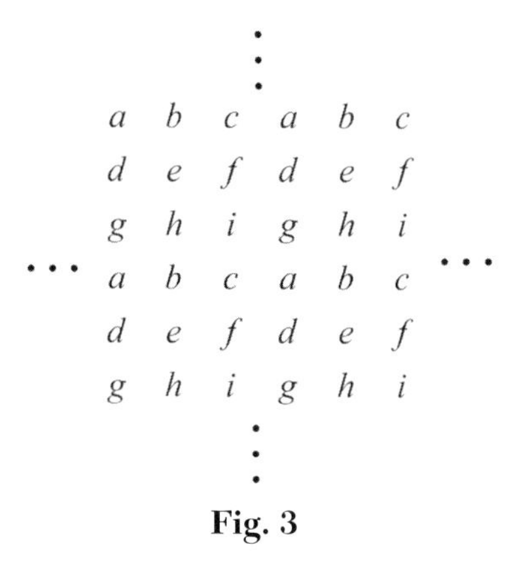

Fig. 3

This same wallpaper pattern can be viewed as if built up from repeated copies of any of its constituent 3×3 areas, of which there are nine, each corresponding to one of the nine possible permutations of the 3 rows and/or 3 columns. All nine are shown in Figure 4.

a	*b*	*c*	*b*	*c*	*a*	*c*	*a*	*b*
d	*e*	*f*	*e*	*f*	*d*	*f*	*d*	*e*
g	*h*	*i*	*h*	*i*	*g*	*i*	*g*	*h*
d	*e*	*f*	*e*	*f*	*d*	*f*	*d*	*e*
g	*h*	*i*	*h*	*i*	*g*	*i*	*g*	*h*
a	*b*	*c*	*b*	*c*	*a*	*c*	*a*	*b*
g	*h*	*i*	*h*	*i*	*g*	*i*	*g*	*h*
a	*b*	*c*	*b*	*c*	*a*	*c*	*a*	*b*
d	*e*	*f*	*e*	*f*	*d*	*f*	*d*	*e*

Fig. 4

Figure 2 is thus one among nine squares that are *indistinguishable* when viewed as toroidally connected. That is, there are nine different ways in which the torus can be cut along two axes perpendicular to each other, unfolded, and then flattened into a square. Or in other words, any particular 3×3 square is but one among nine alternative planar projections of the torus it represents.

Viewed thus, any idea of broken diagonals as the poor cousins in the family disappears, and we discover a simple parity between orthogonals and diagonals, the relations among the 3 rows and 3 columns finding matching counterparts in the relations among the 3 '\'– diagonals, or *slopes* and 3 '/'– diagonals or *slants*. Note however, that the same does not hold of even-order arrays such as 4×4, in which rows and columns intersect on one cell, while slopes (a) and slants (b) may intersect on two; see Figure 5.

a		*b*	
	ab		
b		*a*	
			ab

Fig. 5

A notable consequence of this symmetry in the case of 3×3 arrays is that orthogonal elements can be *switched* with diagonal elements to create a twin or *dual* of the original square as shown in Figure 6.

a	*b*	*c*
d	*e*	*f*
g	*h*	*i*

h	*a*	*f*
c	*e*	*g*
d	*i*	*b*

Fig. 6

The 3 entries found in each orthogonal/diagonal of the left-hand square are the same as the 3 entries found in each diagonal/orthogonal of the right-hand square. It is easy to verify that the squares above are each the *unique* dual of the other, a property not shared by larger odd-order arrays such as 5×5, which admit of alternative possibilities. The transform linking these two squares can then be diagrammed as in Figure 7.

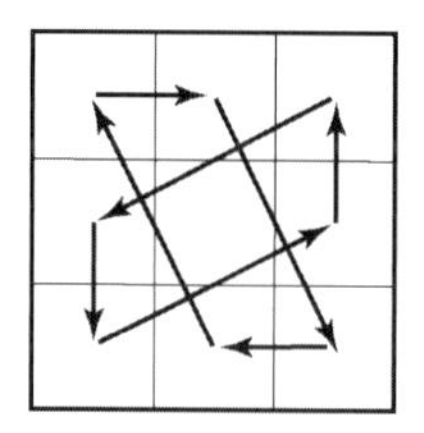

Fig. 7

Elementary as it is, to the best of my knowledge, not a solitary reference to the existence of this duality is to be found in all the literature on magic squares. This is still more amazing in view of its obvious implication, which is that *every 3×3 magic square is but one member of a* ***dyad of complementary squares*** *sharing identical entries, the one magic on all orthogonals and both central diagonals, the other magic on all diagonals and both central orthogonals.*

An immediate question thus prompted is: What then is the complement of the famous *Lo shu*? The answer is *Theon's square*, also known as the 'natural square'.

Little is known about Theon of Smryna, mathematician and astronomer, circa 70–135 BC, whose writings include, ΤΩΝ ΚΑΤΑ ΤΟ ΜΑΦΗΜΑΤΙΚΟΝ ΧΡΗΣΙΜΩΝ ΕΙΣ ΤΗΝ ΠΛΑΤΩΝΟΣ ΑΝΓΝΩΣΙΝ or "Mathematics Useful For Understanding Plato." An 1892 work by J. Dupuis [4] reproduces Theon's original Greek text along with a French translation. The square in question is a 3×3 array showing the first nine letters of the Greek alphabet, one in each cell. It was Greek practice of the period to use letters to stand for numbers. Dupuis' translation of this into arabic numerals is seen in Figure 8 to the right of its dual, the *Lo shu*.

4	9	2
3	5	7
8	1	6

The *Lo shu*

1	4	7
2	5	8
3	6	9

Theon's Square

Fig. 8

It is the square refered to in the opening quotations. Cammann and Ahrens, the authors cited, were of course right to dismiss claims that Theon's square is a magic square. But the fact that there were people who asserted it was, or who saw in it a first attempt at a magic square makes it all the stranger that *nobody* seems ever to have noticed that all six diagonals, as well as both central orthogonals, sum to 15. Least of all, Theon himself, I suspect, who probably knew nothing of magic squares, and who introduces his diagram as a purely didactic device in discussing 5 as the arithmetic mean of pairs summing to 10, the latter a number that looms large in Pythagorean philosophy. In any case, the two statements made about Theon's square, that it is "not a magic square in any sense of the term" and that it has "not the slightest thing to do with magic squares and deserves no place in the history of magic squares" betrays a lack of understanding on the part of these two supposed experts in the field that is simply breathtaking.

Admittedly, from the point of view of the magic square connoisseur, two blemishes mar Theon's square. First, a single glance is enough to grasp its pattern. The serial order of the numbers in the columns makes for a transparent structure that lacks all mystique, and hence charm, and without which it attracts scant interest. Second, unlike the *Lo shu*, in which every straight line of numbers adds to 15, no equivalent elegance is enjoyed by the same-sum triads in Theon's square. However, both blemishes disappear when the two squares are compared *as toroids*, on the surface of which diagonals are no more broken than orthogonals. The dual of a 3×3 magic square is thus not another magic square, but a square of equivalent 'magic.'

There exists a considerable literature devoted to the *Lo shu*, much of it infected with the kind of crypto-mystic twaddle met with in Feng Shui. It would be nice to think that once news of the dual gets out things will change. The discovery that 'the' *Lo shu* is only one-half of a double act is surely going to come as quite a revelation. But of course such a big change will take time to absorb. So the message should be kept simple. How about: "Hey guys, snap out of it: it's not a *lone* shu, it is a *pair* of shus."

References

[1] Shuyler Cammann, The Evolution of Magic Squares in China, *Journal of the American Oriental Society*. LXXX, 1960, pp 116-124.

[2] W. Ahrens, Studien über die magischen Quadrate der Araber, *Der Islam*, VII, 1917, pp 186-250.

[3] L. Sallows, The Lost Theorem, *The Mathematical Intelligencer*, 1997, 19; 4, pp 51-4

[4] J. Dupuis, Theon de Smyrne, *Libraire Hachette et Cie*, Paris 1892, pp.167-8.

Appendix V
The Lost Theorem

Introductory Note

It seems that open season has been declared on the *Lo shu*, because here again in *The Lost Theorem*, I am to be found bemoaning a blemish in its structure. "Why don't you leave that wretched square alone?" I can hear my critics say. But my motive is of course merely practical, for it is only through identifying flaws in existing magic squares that we can hope to produce improved specimens. And the more revered the square in question, then the more searching must be our critical examination. The 'atomic' square here described is a pearl born of the grit at its heart–a particle of grit that is the *Lo shu* itself. Had it not been for defects in the latter, a path to the former might never have been found. The essay appeared in *The Mathematical Intelligencer* in 1997. I am also indebted to Christian Boyer, who has been kind enough to include it on his scholarly and extremely impressive website http://www.multimage.com/indexengl.htm.

The Lost Theorem

"Almost the last word has been said on this subject"
—H. E. Dudeney on magic squares [1]

A magic square, as all the world knows, is a square array of numbers whose sum in any row, column, or main diagonal is the same. So-called "normal" squares are ones in which the numbers used are 1,2,3, and so on, but other numbers may be used. Squares using repeated entries are deemed trivial. We say that a square of size $N \times N$ is of order-N. Clearly, magic squares of order-1 lack glamour, while a moment's thought shows that a square of order-2 cannot be realized using distinct entries. The smallest magic squares of any interest are thus of order-3.

Writing in *Quantum* recently [2], Martin Gardner offered $100 to anyone able to produce a 3×3 magic square composed of any nine distinct square numbers. So far nobody has produced a solution, or proof of its impossibility, although a near miss I discovered is seen in Figure 1, a specimen whose rows, columns, and just one of the two main diagonals sum to the same number, itself a square: 1,472. It was while tinkering in connection with this problem that I was startled to discover an elementary correspondence between 3×3 magic squares and *parallelograms*. The reason for my surprise is worth explaining.

127^2	46^2	58^2
2^2	113^2	94^2
74^2	82^2	97^2

Fig. 1

Magic squares have been a special hobby of mine for over twenty years; the literature on the topic, much of which I have collected, is extensive. As already noted, 3×3 magic squares are the smallest and, hence, simplest types, for which reason they are the earliest to appear in history, as well as being the most thoroughly investigated squares of all. Innumerable books and articles on magic squares begin with a discussion of 3×3 types, the properties of which have long been regarded as completely understood. Writing in the well known *Mathematical Recreations* published in 1930, for instance, Maurice Kraitchik begins by saying that, "The theory of the squares of the third order is simple and complete . . .", and then goes on to present that theory in just two pages of text.

Yet for all its extreme simplicity, the elementary correspondence with parallelograms that I had stumbled upon while working on Gardner's problem, has, to the best of my knowledge, never previously been identified. I feel sure that many readers will share my incredulity on inspecting the theorem below. They may agree with me that the correlation with parallelograms it describes is so basic that it deserves to be regarded as *the* fundamental theorem of order-3 magic squares, and the very first thing that any newcomer to the subject should learn. How then could such a theorem escape the attention of every researcher in the field from ancient times down to the present day?

An explanation lies in the orthodox focus on magic squares using natural numbers. Once our attention broadens to include squares that use *complex* numbers, the familiar integer types become only a special case, preoccupation with which has obscured the wider picture. Moving beyond this narrow view, we step into a realm of greater clarity and harmony. And at the very center of that realm, we shall find an undiscovered prize, the atomic magic square.

Standard Theory

What makes a 3×3 square magic?

Figure 2 shows the well-known algebraic formula due to Édouard Lucas that describes the structure of every magic square of order 3:

$c-b$	$c+a+b$	$c-a$
$c-a+b$	c	$c+a-b$
$c+a$	$c-a-b$	$c+b$

Fig. 2

Lucas's formula conveys much of the essential information in a single swoop. In particular, we can see at a glance that the constant total, which is $3c$, is equal to three times the center number, while a closer look shows that whatever the nine numbers used in the square, they must always include eight 3-term arithmetic progressions, namely:

1 : $c + a, c, c - a,$
2 : $c + b, c, c - b,$
3 : $c + a + b, c, c - a - b,$
4 : $c + a - b, c, c - a + b,$
5 : $c + a - b, c + a, c + a + b$
6 : $c - a + b, c - a, c - a - b,$
7 : $c + a + b, c + b, c - a + b,$
8 : $c + a - b, c - b, c - a - b$

The identification of these arithmetic triads is a recurrent feature in discussions of order-3 theory, a point we shall return to later, although it is rare to find an explicit list of all eight.

Of course, just as any magic square can be rotated and reflected to result in 8 trivially distinct squares that are deemed equivalent, so there are 8 trivially distinct rotations and reflections of the formula, all of them isomorphic to each other, and again comprising one equivalence class.

So much for a bird's eye view of the theory of order-3 magic squares as it is met within the literature. Let us now turn our attention elsewhere.

Complex Squares

Consider Figure 3, which depicts an arbitrary parallelogram, $PQRS$, centered at some arbitrary point, M, on the Euclidean plane, with axes x and y. Point O is the origin of the plane. The corner points, P, Q, R, and S, together with T, U, V, and W, the midpoints of the sides of the parallelogram, as well as the center, M, can thus each be identified with vectors or complex numbers of form $x + y\mathbf{i}$, in which x and y are the real number coordinates of each point, and $\mathbf{i} = \sqrt{-1}$.

Equally, the lines connecting these points may themselves be interpreted as vectors, three of which are identified in the Figure as: $\overrightarrow{MT} = \mathbf{a}$, $\overrightarrow{MU} = \mathbf{b}$, and $\overrightarrow{OM} = \mathbf{c}$. Note that given any three particular complex values for $\mathbf{a}$, $\mathbf{b}$, and $\mathbf{c}$, we could immediately proceed to construct the corresponding parallelogram.

Now, by the law for the addition of vectors, the point or complex number, T (which is the vector $\overrightarrow{OT}$), is the resultant of the two vectors $\mathbf{c}$ and $\mathbf{a}$, or $\mathbf{c} + \mathbf{a}$. And likewise, it takes but a glance to identify the remaining points on $PQRS$ in terms of the vectors, $\mathbf{a}$, $\mathbf{b}$, and $\mathbf{c}$, as indicated in the Figure: $P = \mathbf{c} + \mathbf{a} - \mathbf{b}$, $Q = \mathbf{c} + \mathbf{a} + \mathbf{b}$, $R = \mathbf{c} - \mathbf{a} + \mathbf{b}$, $S = \mathbf{c} - \mathbf{a} - \mathbf{b}$, $U = \mathbf{c} + \mathbf{b}$, $V = \mathbf{c} - \mathbf{a}$, $W = \mathbf{c} - \mathbf{b}$. and $M = \mathbf{c}$.

Looking next at the 3×3 square shown below left in Figure 3, observe what happens when a new square is created by replacing P, Q, . . . with their corresponding expressions in terms of a, b and c, as shown in Figure 4.

The outcome is nothing less than a reappearance of Lucas's formula for 3×3 magic squares.

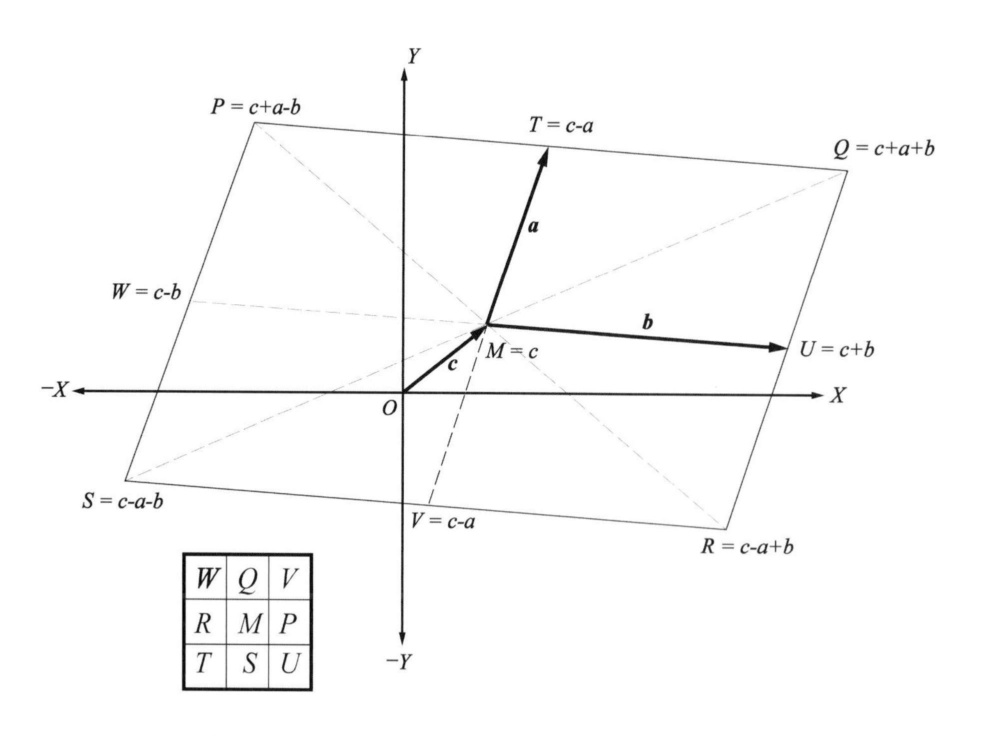

Fig. 3

c − b	c + a + b	c − a
c − a + b	c	c + a − b
c + a	c − a − b	c + b

Fig. 4

The implication is as obvious as it is surprising: given any particular parallelogram on the Euclidean plane, and then transcribing the complex numbers corresponding to its four corners, four edge midpoints, and center, into a 3×3 matrix, in the same way as above, the resulting square will *always* be magic. Or alternatively, starting with any 3×3 magic square that uses complex number entries, we will find that they define nine points on the Euclidean plane that coincide with the four corners, four edge midpoints, and center of a parallelogram.

In summary, we have:

Parallelogram Theorem *To every parallelogram drawn on the plane there corresponds a unique equivalence class of 8 complex 3×3 magic squares, and for every equivalence class of 8 complex 3×3 magic squares there corresponds a unique parallellogram on the plane.*

Or in a nutshell: rotations and reflections disregarded, every parallelogram defines a unique 3×3 magic square, and vice versa.

In this light, it is interesting to recall the eight arithmetic progressions previously identified in every 3×3 magic square. For just as arithmetic progressions of real numbers correspond to equidistant points along the real number line, so arithmetic progressions of complex numbers correspond to equidistant *collinear* points on the plane. See then how the eight progressions listed earlier precisely correlate with the eight sets of 3 collinear points lying along the four edges and four bisectors of the parallelogram in Figure 3:

$1 : c + a, c, c - a = T, M, V$
$2 : c + b, c, c - b = U, M, W$
$3 : c + a + b, c, c - a - b = Q, M, S$
$4 : c + a - b, c, c - a + b = P, M, R$
$5 : c + a - b, c + a, c + a + b = P, T, Q$
$6 : c - a + b, c - a, c - a - b = R, V, S$
$7 : c + a + b, c + b, c - a + b = Q, U, R$
$8 : c + a - b, c - b, c - a - b = P, W, S.$

In the magic square literature to date, discussion of theory never gets further than *identifying* these progressions; now at last we can see how the geometry of the parallellogram *explains* their presence.

From our new perspective we can see also how the corellation with parallelograms has escaped previous notice. Traditionally magic squares have used integers, which are entries without imaginary component. The parallelograms corresponding to these squares are thus collapsed, or *degenerate,* specimens of zero area, making their presence undetectable. In fact, a closer look at one such parallelogram will prove instructive, as well as preparing us for an unexpected development: the discovery of a lost archetype, the primordial magic square.

A Flaw in the Crystal

I suppose that, until now, the most obvious candidate for the title of archetypal magic square would have been the Chinese *Lo shu*, the simplest, oldest, and most well known square of all. It is seen in Figure 5.

2	9	4
7	5	3
6	1	8

Fig. 5

Legend has it that the *Lo shu* was first espied by King Yü on the back of a sacred turtle that emerged from the river Lo in the 23rd century BC. In fact, historical references to the square date from the 4th century BC, while Cammann has argued that it played a major part in Chinese philosophical and religious thought for centuries afterward [3]. In the West, the *Lo shu* has long been held up as a paradigm, or "one of the most elegant patterns in the history of combinatorial number theory," as Martin Gardner has written. Nevertheless, taking a lens to this ancient gem, we can discover an interesting irregularity in its crystal lattice.

Consider the *Lo shu*'s flattened parallelogram, which is that segment of the real number line between 1 and 9, along which lie its four corners, four edge midpoints, and center, occupying nine equidistant points. Recalling Figure 3, the relation between these points and their position in the magic square can be diagrammed as in Figure 6.

The distance between the parallelogram's corners at points 1 and 3 (or 7 and 9) is thus 2 units, while the distance between those at points 3 and 9 (or 1 and 7) is 6

```
      3     6     9                2 9 4
   2     5     8          ⇔        7 5 3
1     4     7                      6 1 8
```

Parallelogram **Magic Square**

Fig. 6

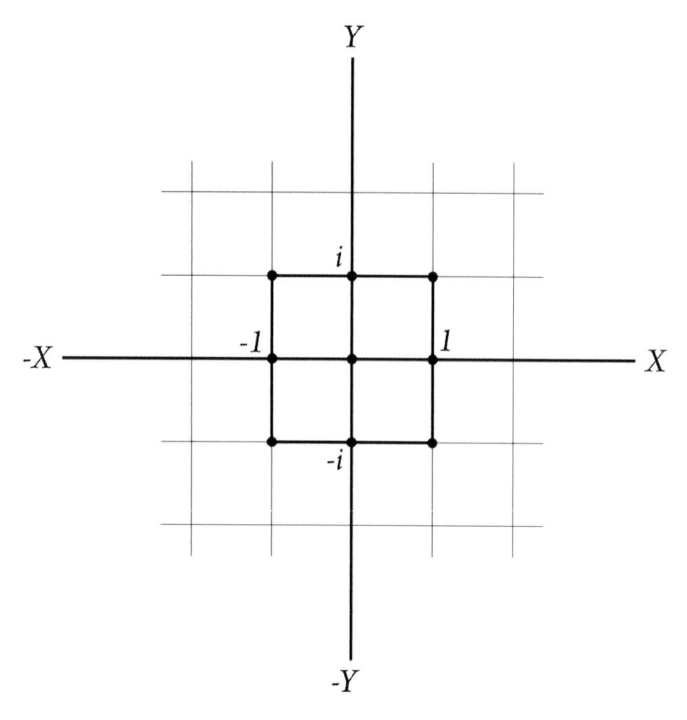

Fig. 7

units; a ratio in side lengths of 1:3. The remarkable fact is thus that the *Lo shu* parallelogram is not equilateral, which is a bit disappointing for a pattern whose famous *symmetry* has won acclaim down the ages, from the banks of the river Lo to the pages of *Scientific American*. It is beginning to look as if that turtle was not quite as sacred as King Yü had imagined.

However, as a prisoner in Flatland, the *Lo shu* is doomed to this imbalance: squash any *equilateral* parallelogram, and two pairs of edge midpoints and two corners will *coincide*, forcing repeated entries in the associated magic square. In other words, unless it is trivial, an asymmetrical parallelogram must accompany *every* non-complex magic square, the one on the (mock?) turtle included.

Where then is the magic square with the symmetrical parallelogram that King Yü was denied?

The Atomic Square

Of course, the most symmetrical case of all is an equilateral parallelogram that is equiangular as well: the *square*.

A magic square whose associated parallelogram is again a square; the idea is at once compelling. But what kind of a magic square would that be? To find out, all we have to do is draw a square on the plane, read off the complex values of its four corners, etc, and then write these into a 3×3 matrix in the usual way. Simplest of all is the canonical or atomic case seen in Figure 7.

It is a square centered on the origin of the plane, such that its four corners and four edge midpoints coincide with the 8 complex integers immediately surrounding the origin. The magic square corresponding to this geometric square is consequently an atomic paradigm of its kind too: it is the smallest, most perfectly symmetrical magic square, composed of the nine smallest Gaussian integers:

$-\mathbf{i}$	$1+\mathbf{i}$	-1
$-1+\mathbf{i}$	0	$1-\mathbf{i}$
1	$-1-\mathbf{i}$	$\mathbf{i}$

Fig. 8

The elegance of this flawless prism is beyond compare. The two main diagonals and two central orthogonals are like four balanced beams pivoted on the center number, the integer at the end of each beam offset by its opposing negative image, an equipoise reflected in the magic sum of zero. Rewriting the square in the form of vectors as ordered pairs, $[a, b]$, its structural harmony reappears in the shape of palindromic rows and antipalindromic columns, a quality that is better highlighted when $\overline{1}$ replaces -1, and the commas and brackets are discarded: see Figure 9.

Analysing the square in terms of Lucas's formula, we find that the variables a and b have here taken on the

$0\bar{1}$	11	$\bar{1}0$
$\bar{1}1$	00	$1\bar{1}$
10	$\overline{11}$	01

Fig. 9

values of 1 and **i**, the real and imaginary forms of unity, while c is equal to zero. Could anything be more natural, or poetic?

My interest in magic squares began a couple of decades ago when I first encountered the *Lo shu*. I recall my delight in exploring its symmetries, but I also recall my disquiet in detecting a strange lopsidedness. In Lucas's formula, the variables a and b appear in two patterns that are perfect mirror images. In the *Lo shu*, however, $a = 1$, while $b = 3$, a numerical imbalance that clashed with the symmetry of the patterns. Attempts to construct a square in which $a = b$, or $a = -b$, wouldn't work either, because the result is then trivial. Yet a craving for symmetry is what makes the mathemagical mind tick. Down the years this unease has continued to quietly smoulder, until recent events brought the parallelogram theorem to light, and, with it, a sudden resolution of the mystery in the shape of the atomic square, whose symmetry is without flaw. It is a relief; I look forward to sleeping at nights once again.

The above article first appeared in *The Mathematical Intelligencer* Vol 19, No. 4, pp 51-4, 1997. It is reprinted here with the kind permission of Springer Science + Business Media.

References

[1] H.E. Dudeney, *Amusement in Mathematics*, p. 119.

[2] M. Gardner, The Magic of 3×3, *Quantum*, January/February 1996, pp. 24-6.

[3] S. Cammann, The Magic Square of Three in Old Chinese Philosophy and Religion, *History of Religions I* (1961), pp. 37-80.

Glossary

alphamagic square A special kind of numerical magic square, the numbers of which can be represented by their word equivalents ('one,' 'two,' etc.) so that the total number of *letters* then occurring in each row, column and diagonal, is the same.

algebraic generalization An array in which algebraic terms consisting of variables and operators describe the necessary relations among numbers occupying the cells of a corresponding magic square.

almost-magic square A square in which every row and column, as well as just one of the two main diagonals is magic.

ambimagic square A novel kind of magic square (due to the author) in which the sum of the numbers in every orthogonal (row and column) is the same, while the product of the numbers in every diagonal (including 'broken' diagonals) is also constant. It can be shown that the magic sum in an ambimagic square of order-3 is necessarily zero. The converse of an ambimagic square (constant orthogonal products, constant diagonal sums) is a 'mabimagic' square.

area square The numerical magic square that is formed by the piece areas in a corresponding 2-D square. Since differently shaped pieces may be of same area, area squares are frequently trivial.

bi-magic square In the realm of numerical magic squares a bi-magic specimen is one in which the squares of the numbers appearing also form a magic square. In the present account, the term has been borrowed to indicate a geomagic square in which the pieces are able to tile *two* distinct targets.

co-diagonal In any square, the diagonal running from top right to bottom left (/).

compact A (geo)magic square of 4×4 in which the four elements in each of its sixteen 2×2 sub-squares will combine to produce the constant total or target. Every nasik numagic square of order 4 is compact, and every compact numagic square of order 4 is nasik. 2-D squares can however be nasik without being compact, or compact without being nasik.

complementary pairs A numerical magic square of order N is composed of $N^2/2$ complementary pairs when the latter all sum to half the magic constant. Similarly, a *geometrical* magic square of order N is composed of $N^2/2$ geometrically complementary pairs when the latter will combine to produce an identical shape. See geometric *complement*.

degenerate square A square containing repeated entries, also known as a *trivial square*.

demi-nasik A geomagic square in which, besides the two main diagonals, there exist at least two non-parallel broken diagonals that are also magic (i.e., contain target-tiling piece sets). See *semi-nasik*.

detrivialize In the case of algebraic squares, the process of adding variables so as to produce a unique expression in every cell. In the case of geometric squares, the process of appending/excising keys and keyholes so as to yield a distinctly shaped piece in every cell.

dimension The dimension of a geomagic square is simply the same as that of the pieces of which it is comprised.

disconnected piece A compound piece consisting of two or more separated islands considered as a single structure whose components are joined rigidly, if invisibly, to each other; also, a *disjoint piece*.

disjoint piece See *disconnected piece*.

disjoint arcs A particular case of a disconnected piece in which the separated components are all segments belonging to the same circular arc.

empty piece The invisible man among pieces. A piece consisting of an empty set of points.

Eulerian square A square formed by 'superimposing' two suitable Latin squares such that every cell becomes occupied by a distinct entry. Also known as a *Graeco-Latin square.*

diagonal Latin square A latin square in which the set of entries occupying each of the two main diagonals is the same as it is for each row and column.

Dudeney's graphic types A classification of the 880 normal order-4 magic squares into twelve distinct Types, according to the distribution of their constituent complementary pairs.

fertility The number of different magic squares that can be produced using a given set of N^2 distinct numbers.

five types of area square The five possible algebraic structures underlying numerical magic squares, whether containing repeated entries or not.

formula An alternative expression for an *algebraic generalization.*

geo-alphamagic square A cross-breed between a geomagic square that uses polyominoes and an alphamagic square. Distinct number-words adorn the polyomino-pieces, a single letter appearing in each cell. The numbers are so chosen that their sum is the same in every line, just as the piece shapes are chosen so as to tile the target in every line.

geo-Eulerian Square A geometrical interpretation of a Graeco-Latin or Eulerian square.

geomagic square A square array of N-dimensional figures that will combine in every line so as to form an identical compound figure. See the formal definition in Appendix 1.

geometric complement A set of pieces is said to comprise so many complementary pairs when the two members of each pair will combine so as to form an identical shape. The piece pairs are then said to be the geometric complement of each other.

geometric variable On analogy with algebraic variables, a certain shape may be regarded as a 'geometric variable', in that it can be seen as standing for a range of contingent shapes.

geolatin square A Latin square of $N \times N$ using geometrical shapes as elements, and such that the N shapes in each line tile a common target. Alternatively, a trivial geomagic square derived from a Latin template.

Graeco-Latin square See *eulerian square.*

graphic types See Dudeney's graphic types.

isometric grid A regular grid like a rectangular grid that is seen on tiling the plane with unit equilateral triangles joined edge to edge.

key A shape that is appended to an existing piece

keyhole An indentation or region that is excised from an existing piece

Latin square An $N \times N$ array containing N distinct entries, each of them occurring exactly once in every row and column.

Lo shu A famous 3×3 magic square of early Chinese origin.

lug-type key A shape that is appended toa piece so as to form a distinct projection, in contrast to the result of appending a *size-altering key.*

mabimagic square See *ambimagic square.*

magic line Short for a row, column, diagonal or other group of cells, the entries of which will combine to produce the required constant outcome (add to required sum or tile required target).

main diagonal In any square, the diagonal running from top left to bottom right (\).

nasik A magic square in which every diagonal, including the so-called 'broken' diagonals, is magic. Nasik squares are also known as panmagic, pandiagonal, diabolic and satanic.

normal square A numerical magic square of size $N \times N$ that uses the numbers 1, 2, . . . , N^2. A 2-D geomagic square whose area square is a normal numagic square.

numagic square An abbreviation for 'numerical magic square.'

order The size of a magic square. A square of size $N \times N$ is said to be of order-N.

orthogonals The rows and columns of a square.

pentomino A polyomino composed of five unit squares.

picture-preserving square A geomagic square in which the surface composition of each target is the same.

polycube A three-dimensional shape formed of unit cubes that are joined face-to-face.

polyform A figure, shape, or structure composed of repeated atoms. The latter may be of any dimension, provided all are the same.

polymagic square A geomagic square in which the pieces used are all polyforms.

polyomino A shape composed of unit squares joined edge -to-edge.

polyiamond A shape composed of unit equilateral triangles joined edge-to-edge.

polyhex A shape composed of unit regular hexagons joined edge-to-edge.

toroidally-connected A square regarded as if inscribed on a torus. That is, the square is imagined to be rolled into a cylinder which is then bent smoothly in a circle until its two ends meet. The top row thus becomes adjacent to the bottom row, and the left-hand column adjacent to the right-hand column. What appear as broken diagonals in the original square now become fully-fledged diagonals.

self-interlocking square A special kind of geomagic square using essentially square pieces that can be made to coincide with the square-cell boundaries of the array that contains them. The effect produced is spectacular. Thus far, the only known specimens are of order-4. Whether or not there exist self-interlocking squares of higher orders is unknown.

semimagic square A square in which only rows and columns are magic. An orthogonally magic square.

semi-nasik A (geo)magic square in which, besides the two main diagonals, there exist at least two *parallel* broken diagonals that are also magic (i.e., contain target-tiling piece sets). See *demi-nasik*.

size-alterning key A shape that can be appended to an existing piece so as to produce an apparent enlargement or elongation of the piece, rather than the suggestion of a distinct appendage.

substrate A trivial geomagic square showing repeated piece shapes that is used as a starting point for further elaboration into a non-trivial square.

target The common shape, region, or structure that is formed or tiled by assembling its component pieces that are found occupying every row, column, and main diagonal in a geomagic square.

template An algebraic square that is used as a guide in constructing a geomagic square. In doing so, distinct variables are translated into distinct geometric shapes. Where variables occur in both positive and negative form, the former are interpreted as shapes to be appended, the latter as shapes to be removed or cut out.

three-dimensional square A geomagic square using solid rather than planar pieces. In general, geomagic square may employ pieces of any dimension.

trivial square A square in which one or more entries occur more than once.

uniform array A square in which every entry is the same.

weakly-connected Pieces which are almost disconnected but in which some component regions are not wholly separate but meet in a single point.

References

[1] S. Cammann, "Old Chinese magic squares," *Sinologica*, 7(1962): 14-53.

[2] S. Cammann, "The magic square of three in old Chinese philosophy and religion," *History of Religions* 1 (1961): 37-80.

[3] L. Sallows, Alphamagic squares, in Richard Guy & Bob Woodrow, *The Lighter Side of Mathematics*, (Math. Assoc. of America, 1994), pp. 305-339.

[4] L. Sallows, "The Lost Theorem," *The Mathematical Intelligencer*, 19:4 (1997): 51-54.

[5] Édouard Lucas, *Recreations Mathematiques*, Vol. IV (1894): 225.

[6] M. D. Hirschhorn The 1976 Summer School, *Parabola*, Vol 13 Issue 1 2-5,18; Issue 2: "More Tessellations with Convex Equilateral Pentagons," 20-2.

[7] M. D. Hirschhorn; D. C. Hunt, "Equilateral convex pentagons which tile the plane," J. Combin. *Theory Ser.* A, 39 (1985): 1-18.

[8] Martin Gardner, *Time Travel and Other Mathematical Bewilderments*, (San Francisco: W.H. Freeman and Co., 1988), pp. 174-5.

[9] Doris Schattschneider, *In Praise of Amateurs*, ed. David A. Klarner, (Prindle, Weber, Schmidt 1981), pp. 140-66.

[10] L. Sallows, "Geomagische vierkanten," *Pythagoras* 1 (Sept. 2009): 12-15.
L. Sallows "Geomagische vierkanten met Lucas' formule," *Pythagoras*, 2 (Nov. 2009).
L. Sallows "Panmagisch versus most perfect," *Pythagoras* 4 (Feb. 2010).
L. Sallows "4×4 Geomagisch vierkant met 8 complementaire paren, *Pythagoras* 1 (Sept. 2010).

[11] J. Dénes & A.D. Keedwell, *Latin Squares and Their Applications*, (English Universities Press, 1974).

[12] H.E. Dudeney, *Amusements in Mathematics*, (New York: Dover Publications, Inc.,1958).

[13] Barclay Rosser and R. J. Walker, "On the transformation group for diabolic magic squares of order four," *Bulletin of the Mathematical Society* Vol XLVI (June 1938): 416-20.

[14] L.S. and Penrose R., *British Journal of Psychology* 49 (1958): 31-3.

[15] M. Coster, "Prijsvraag: wees een magiër," *Pythagoras* 2 (Nov. 2009): 14-17.

[16] M. Coster, "De magiërs zijn onder ons," *Pythagoras* 6 (June 2010): 16-19.

[17] Martin Gardner, "Mathematical Games," *Scientific American* Vol 234 No 1 (Jan. 1976): 118-223.

[18] Greg Ross, http://www.futilitycloset.com/index.php?s=lee+sallows

[19] L. Sallows, "Nieuw archief voor wiskunde" *Vijfde Serie* deel 10, No 1(Maart 2009): 70.

[20] Jack Chernick, "Solution of the general magic square," *American Mathematical Monthly* Vol 45 (1938): 172-5.

[21] H.S. Hahn, "Another property of magic squares," *Fibonacci Quarterly* 13 (Oct. 1975): pp 205-8.